# Construction Contract Administration for Project Owners

*Construction Contract Administration for Project Owners* is aimed at public and private owners of real estate and construction projects. The book is intended to assist owners in their contractual dealings with their designers and their contractors. Most owners are not primarily in the business of designing and building facilities. The fact that their primary business is not design and construction places them at a disadvantage when negotiating, drafting, and administering design agreements and construction contracts because their designers and contractors use these documents every day. This book is intended to assist owners to redress this imbalance by equipping owners to draft and administer contracts so as to protect their interests.

The book is aimed at owner personnel with all levels of knowledge in the business of managing projects. It can serve as a comprehensive introduction to drafting and administering design agreements and construction contracts for beginners. For intermediate level personnel, it can serve as a manual to be read to enhance the reader's skills in this area. For the sophisticated project management professional, it can serve as a resource to be consulted in connection with very specific issues as they arise on a project.

**Claude G. Lancome** is Executive Vice President and Principal in Coast and Harbor Associates, a firm specializing in assisting public and private owners to manage the design and construction of their facilities. His experience includes projects with an aggregate value in excess of $5 billion. Mr. Lancome is a graduate of Earlham College and Harvard Law School.

# Construction Contract Administration for Project Owners

Claude G. Lancome

Routledge
Taylor & Francis Group

LONDON AND NEW YORK

First published 2017
by Routledge
2 Park Square, Milton Park, Abingdon, Oxon OX14 4RN

and by Routledge
711 Third Avenue, New York, NY 10017

*Routledge is an imprint of the Taylor & Francis Group, an informa business*

*British Library Cataloguing-in-Publication Data*
A catalogue record for this book is available from the British Library

*Library of Congress Cataloging in Publication Data*
Names: Lancome, Claude G., author.
Title: Construction contract administration for project owners / Claude G. Lancome.
Description: Abingdon, Oxon [UK]; New York : Routledge, 2017. | Includes bibliographical references and index.
Identifiers: LCCN 2016038862 | ISBN 9781138244252 (hardback : alk. paper) | ISBN 9781315277004 (ebook : alk. paper)
Subjects: LCSH: Construction contracts. | Construction contracts—Management. | Contracts (International law)
Classification: LCC K891. B8 L36 2017 | DDC 343.07/8624—dc23
LC record available at https://lccn.loc.gov/2016038862

ISBN: 978-1-138-24425-2 (hbk)
ISBN: 978-1-315-27700-4 (ebk)

Typeset in Times New Roman
by Keystroke, Neville Lodge, Tettenhall, Wolverhampton

# Contents

# Introduction

This book is about successful contract management for owners of facilities projects. Successful contract management involves both effective contract drafting and effective contract administration. The book covers both topics: how to draft, and how to administer design agreements and construction contracts in order to achieve the owner's objectives while protecting the owner's interests.

The book is based on the author's involvement in contract management on projects which had a total value of more than $5 billion. Individually, these projects ranged in size from $500,000 to $3.5 billion (consisting of more than 100 separate contracts). In each instance, the author worked on behalf of the project owner as a member of the owner's representative team.

The author's experience working for owners has led to one basic conclusion. Most owners—whether public, private, or nonprofit—are not nearly as experienced, skillful, and interested in managing the owner–designer and owner–contractor relationships as are their designers and their contractors. This mismatch arises from several factors: some general, and some specific to individual owners. The most important reason is that, for most owners, managing relationships with designers and contractors is ancillary to the primary mission of the organization; whereas, for designers and contractors, managing and benefiting from relationships with owners is the primary mission of the organization.

The purpose of this book is to assist owners to eliminate any disadvantage they may have in managing contractual relationships with their designers and contractors. The goal is to position owners to manage these relationships successfully. The book also addresses relationships with consultants. These are consultants that the owner elects to contract with directly (as opposed to consultants that are subconsultants to the lead designer).

The relationship between the owner and its designer, and the relationship between the owner and its contractor, are established and defined by the contracts, respectively, between the owner and the designer and between the owner and the contractor. That means that for the owner to successfully manage these relationships, it must draft and administer these contracts effectively. The rights and responsibilities of the owner and the designer are established by the design agreement. Similarly, the rights and responsibilities of the owner and the contractor are established by the construction contract. These documents create "the law of the project." All issues that arise in the course of the project must be resolved as required by these documents. Understanding these basic business and legal realities is the first step to understanding effective contract management.

The designer's scope of service, its standard of care, and its entitlement to additional compensation are all governed by what is in the design agreement. Similarly, the construction contract determines the contractor's scope of work, its administrative responsibilities with respect to managing the project, and its entitlement to additional compensation and/or additional time in which to perform the work.

Some designers and contractors will try to resolve issues on the basis of external factors. They will refer to such things as industry practice ("Everybody does it that way around here."), level of effort ("We're working real hard to give you a great project."), and project economics ("We're losing our shirt on this job.") as bases for determining the appropriate resolution of an issue. The experienced contract administrator knows that what matters is what is in the contract; the external factors are, in most cases, irrelevant. This book will discuss how to use the design agreement, consultant agreement, and the construction contract to resolve issues.

Since these contracts determine how the project will be designed and built, and how issues will be resolved, it is very important for the owner to consider carefully the design agreement, any consultant agreements, and the construction contract prior to beginning the project. An owner is significantly more likely to effectively administer a design or consultant agreement, or construction contract, if the document is drafted with the owner's interests in mind. This is why owners that do repetitive construction should create a standard owner–designer agreement, a standard owner–consultant agreement, and a standard owner–contractor contract. Such documents will ensure that the owner's interests are protected in all the owner's projects. This book offers sample provisions and examples of a complete design agreement, consultant agreement, and construction contract to assist owners to develop their standard documents.

The book addresses effective drafting and administration of owner–designer agreements and owner–contractor contracts on two levels. On one level, the book is a "how-to" manual for owners on drafting and administering contracts. There are numerous specific recommendations. The reason, or reasons, for the specific approach is explained and an example or sample is provided. On another level, the book is about empowerment. It shows owner personnel that they can be as efficacious on behalf of their organizations as designer, consultant, and contractor personnel are on behalf of theirs. This is accomplished through explanations of strategies that provide support and guidance for the specific activities discussed throughout the book.

Owner staff at all levels should find this book helpful. For the beginner, the book serves as a soup-to-nuts guide on strategies and techniques. The intermediate reader can use the book to sharpen skills, learn new techniques, and develop a more strategic approach to contract administration. For the expert, the book provides an as-needed resource with new ideas on how to most effectively draft and administer design and consultant agreements, and construction contracts, and on how to resolve any of the myriad of contract administration issues that may arise in the course of designing and building a facility.

Having described the essential purpose and goal of the book, it may be useful to briefly indicate how this book differs from other books on design and construction projects. It is not about project management; it is about managing the legal and business relationships that are at the heart of each project. It is not about how to run an owner's organization; it is about assisting members of the owner's organization to more effectively identify and resolve issues and disputes in their best interests. It is not an attempt to present information to owners, designers, and contractors in a balanced manner; it is unequivocally devoted to presenting concepts and information to owners so they can be more effective in their dealings with designers, consultants, and contractors. And it is not a book for current or prospective homeowners interested in building or rehabilitating their home; it is for personnel who work for owner organizations-business firms, government agencies, and nonprofit entities. Indeed, the book is intended to be useful to all the staff of an organization that are involved directly or indirectly with design and construction, ranging from presidents and executive directors, to chief financial officers, to directors of real estate and construction, to inside and outside counsel, to project managers, and to anyone else involved in the design and construction process in any organization.

Since in the vast majority of projects the owner relies on the designer for construction administration, that is the model assumed here. That is, it is assumed that there are three parties involved in the project: the owner, the designer, and the contractor. The owner needs to be actively involved in managing its relationship with its designer and its contractor. That is because, as owner, one has a unique level of interest in scope, cost, and schedule that just aren't shared by the other project participants. To put it another way, the owner does not have the same interests as its designer and its contractor.

An owner who fails to get actively involved in drafting and administering the design agreement and the construction contract is essentially relying on the designer and the contractor to determine the scope of the project, control costs, and protect the schedule. Reading and then thinking about this statement should make it crystal clear to owner personnel why their active involvement in contract management is critical.

This book is aimed at assisting owner personnel to get actively involved and to do so effectively. It is about how to control scope, costs, and schedule through effective contract management. Contract management is only one part of project management, and, therefore only one aspect of addressing scope, cost, and schedule. But it is a critical, often overlooked, part of project management, and one with potentially significant payoff to the owner.

This book discusses the drafting and administering of contracts. It is intended as a guide, not a substitute for the advice of an attorney who is familiar with design and construction law in the jurisdiction of the project. On the other hand, construction contract attorneys typically do not have experience working on actual projects seeing how contract provisions do and don't work. It is very expensive to use an attorney to advise on contract administration issues, and, in the author's experience, not many attorneys in private practice understand the practical aspects of administering contracts. For that reason, this book is intended to provide sufficient in-depth advice to equip the typical public and private owner with the ability to effectively draft and administer design and consultant agreements and construction contracts on their own, limiting formal legal advice to a final review of contracts if necessary and to resolving disputes that cannot be resolved by the project participants.

We are now ready to briefly summarize the contents of the book so that the reader may have a somewhat better understanding of what lies ahead. The book is divided into three parts. The first part, consisting of Chapters One through Three, discusses contract drafting and contract administration in general terms. The second part, consisting of Chapters Four through Six, addresses the drafting and administration of design and consultant agreements. The final part, consisting of Chapters Seven through Ten, addresses the drafting and administration of construction contracts.

Chapter One discusses project delivery methods. It is important to decide on the project delivery method before drafting the project agreements. That is because different project delivery methods will lead to variations in certain contract provisions.

Chapter Two presents a general description of how to draft effective contracts. This discussion applies to design agreements, consultant agreements, and construction contracts. After completing this chapter, the reader will understand the principles of drafting effective contracts.

Chapter Three introduces contract administration. It starts by discussing contract deliverables and other products of the contract administration process. It then explains techniques for effectively administering contracts. These techniques apply to design and consultant agreements and construction contracts. After completing this chapter, the reader will have an in-depth understanding of what contract administration is and how to use a number of techniques to achieve positive results for the owner.

Chapter Four addresses how to draft the owner–designer agreement so as to protect the interests of the owner. A sample owner–designer agreement is included in the Appendix at the end of the book as Appendix A. Chapter Five explains how to administer this agreement to provide the best results for the owner. It focuses on how to obtain those products and activities that were promised by the designer in the design agreement. It then addresses additional services, which is designer terminology for additional compensation. The focus is on how to minimize these additional costs. Chapter Five also addresses the issue of designer responsibility for errors and omissions and explains how and to what extent the owner can hold its designer financially responsible for its mistakes. Chapter Six addresses consultant agreements. Owners may use consultants with whom they wish to contract directly as part of the project team. This chapter addresses drafting consultant agreements and provides a sample agreement, which is included in Appendix B. After completing these chapters, the reader will have important insight into how to draft and administer the owner–designer agreement and the owner–consultant agreement so as to most effectively protect the owner's interests.

Chapter Seven discusses how to draft the construction contract in the owner's best interests. A sample contract is provided in Appendix C. Chapter Eight deals with administering the construction contract. Chapter Nine addresses change orders and claims. Since there are many books that have been written about claims, this chapter does not seek to duplicate all this material in one chapter. There is a review of the most common types of change order proposals and the contractor's burden of proof for each of these types of proposals. The emphasis in this chapter is on how owners can most effectively defend against change order proposals. After completing these three chapters, the reader will have a good understanding of how to draft and administer the construction contract, including how best to defend against the most common types of change order proposals.

Chapter Ten addresses closing out the project. This is a very important part of successfully completing a project because it is the process through which the owner assures that the contractor has completed all its obligations under the contract. However, because it is difficult, and because most participants are in a hurry to get to the next project or to use the facility, it does not get the attention it deserves. This chapter covers administering the punch list, the importance of substantial and final completion, documenting the project's completion, and determining final payment. After completing this chapter, the reader should have an understanding of how to effectively administer the completion of the contract to achieve the owner's objectives and protect the owner's resources.

Having stressed how critical contract management is, it is nonetheless important to remember that contract management is only one part of project management. The scope of project management is the management of the entire project. This involves obtaining the necessary financing, management of design and construction, budgeting and cost reporting, scheduling, estimating, managing relationships with other parts of the owner organization and with government agencies and community groups with an interest in the project, managing relationships with the user organization (whether internal or external), and all other aspects of managing the design and construction of a facility. Contract management is only one part of project management, but a critical one because of its potential impact on scope, cost, and schedule.

In summary, this book is about how owners can more effectively draft and administer the owner–designer agreement, the owner–consultant agreement, and the owner–contractor contract. It addresses this topic on both a strategic and on a how-to level.

# 1 Project delivery methods

## Introduction

Before beginning to draft either the design agreement or the construction contract, the owner should carefully consider its project delivery method. Each method offers certain advantages and challenges to the owner relative to other methods. Because there are a number of available books that address alternative procurement methods in detail, the descriptions in this chapter will briefly summarize the major attributes of each method. The major focus will be the extent to which each method benefits the owner. The chapter includes a discussion of project delivery methods that aim at enhanced collaboration between project participants.

Before discussing each of the specific delivery methods, it is important to discuss one of the most important elements in determining whether an owner–designer and an owner–contractor relationship will work productively. This element is closely associated with the procurement process, and, therefore, it is discussed in this chapter.

## Relationship as the basis of procurement

One very key (but by no means the only) determinant of the productivity of a relationship between the owner and the designer, and between the owner and the contractor, is the existence or potential existence of an ongoing, mutually beneficial relationship. It is the ongoing nature of the relationship that is particularly important.

If the owner believes the designer or contractor has done good work for it in the past and is doing good work now, the owner will want the designer or contractor to do more work for it in the future. For that reason, the owner will be flexible when considering proposal or bid values, contract administration issues, and proposed change orders. If the designer or contractor believes the owner has accepted reasonable proposal or bid values, has been reasonable on change order and other issues, and has paid dependably, the designer or contractor will want to do more work for the owner. For that reason, the designer or contractor will be less aggressive when submitting proposals or bids, when preparing change order proposals, and when addressing contract administration issues.

For this type of ongoing relationship to exist and to be the basis of a productive relationship, three conditions must exist. First, the owner has to regularly design and build facilities. Second, the owner has to be able to award design and construction work on the basis of criteria other than, or in addition to, price. Most important is the ability to consider the owner's previous experience with a design or a construction firm. That is the criterion that allows owners to reject proposals or bids from firms that, in the owner's opinion, were not reasonable on prior projects. This, in turn, creates a strong incentive for designers and contractors to be reasonable in their approach to proposing or bidding, pricing change orders, and addressing other issues. The third necessary condition is that the personnel responsible for procuring and delivering the project for the owner, the designer, and the contractor, understand the value of a relationship.

The absence of any of these conditions makes it somewhere between very difficult and impossible to derive the benefits of the actual or potential relationship. If the owner only designs and builds a facility once in a great while (say, every five years or so), the interaction between the owner and the designer, and the owner and the contractor, is not frequent enough for any of the participants to consider it a relationship. If the owner can only award on the basis of price, as is the case for many public owners, then the prior experience element cannot be considered and that part of the incentive is removed for the designer and the contractor. If the personnel working for the owner, the designer, and/or the contractor do not value the relationship, then it is very difficult to derive the benefits because decision-making will be made without reference to the relationship.

The importance of a relationship applies to consultants as well. However, these agreements typically involve a very specific scope of service and much smaller agreement values so the importance of the relationship is not as prominent as it is for designers and contractors.

## Project delivery methods

The key role of the relationship applies to all the project delivery methods. There are six common project delivery methods that are listed in Exhibit 1.1.

---

**Exhibit 1.1   Most common methods of project delivery**

- Design-bid-build.
- Design-build.
- Construction manager at risk.
- Fast track.
- Methods seeking enhanced collaboration (LEAN Construction and Integrated Project Delivery).
- Project management consultants.

---

### *Design-bid-build*

This method involves selecting a designer who prepares the plans and specifications that fully describe the project to be built. When the contract documents are complete, the project is put out to bid. Contractors interested in building the project submit bids, and the contract is awarded to the qualified bidder that submits the lowest bid.

Design-bid-build provides owners two important benefits. First, because the documents are complete before bids are submitted, the winning bid should reflect reasonably accurately the cost to build the project. To put it another way, because, other than in the case of unforeseen site conditions, there should be no significant unknown aspects of the project, meaning that the winning bid is based on full knowledge of the project, and, therefore, should be reasonably accurate. Also, because all factors are known, it will be difficult for the contractor to justify change orders that do not represent owner-directed changes.

The second advantage for the owner is the checks and balances provided by the design-bid-build approach. The owner has a direct contractual relationship with both the designer and the contractor. They both owe the owner the benefits for which the owner has bargained in each of those contracts. Furthermore, the designer typically has significant contract administration responsibilities which can assist the owner in achieving a successful project.

There are two significant disadvantages that can occur. First, the contractor and the designer can have disputes about what caused problems on the project, each seeking to avoid responsibility. Second, some designers are not skillful at all aspects of contract administration, thereby giving the upper hand to the contractor. This can cause adverse financial consequences for the owner.

### Design-build

This method involves procuring one entity to provide both design and construction services. The design may be completed before construction starts, or construction may start before the design is completed.

From the owner's standpoint, there are two significant advantages to the design-build approach. First, there is a single point of contact and responsibility. The owner has one contractual relationship to manage: the one with the design-build entity.

Either the designer or the contractor can be the lead organization for a design-build team. There is no inherent advantage to the owner in either arrangement. The best team will depend on the strength of the team members in relationship to the needs of the particular project. There are also companies that offer both design and construction services. If the company provides quality services, to have both design capabilities and construction capabilities in the same company can provide further efficiencies in management for the owner.

The second advantage to the owner of design-build is the elimination of conflicts between the designer and the contractor. When problems arise on a project, there are no longer any incentives for the designer and the contractor to point fingers at each other. They are now on the same team and share an incentive to avoid problems.

This points to the major disadvantage for the owner in the design-build approach. There is no longer any check and balance between the designer and the contractor. While the designer and the contractor now share an incentive to avoid problems, they also share an incentive to assign responsibility to the owner for all problems that cannot be avoided. Furthermore, contract administration on behalf of the owner is now performed solely by the owner. Such activities as reviewing requisitions for payment, responding to submittals, and evaluating change order proposals are done exclusively by the owner because the designer is "on the other side."

### Construction manager at risk

There are two types of construction manager. This section discusses construction manager at risk. The section on project management firms discusses the other type of construction manager, construction manager as agent.

There are several differences between the construction manager at risk and the general contractor. It is these differences that offer the advantages to the owner compared to a general contractor.

The major conceptual difference is that a construction manager is supposed to operate on behalf of the owner, as opposed to a general contractor who is supposed to deliver the facility.

Construction management contracts typically refer to the relationship between the construction manager and the owner as "a relationship of trust" or similar wording. The construction manager is supposed to protect the interests of the owner while building the facility, whereas the general contractor is only obligated to complete the facility.

The second difference is that a construction manager provides preconstruction services. These include such services as constructability reviews of design documents, value engineering, cost estimating, and construction planning. The construction manager is supposed to work cooperatively with the owner and the designer to develop the best possible project for the owner.

The third difference is that the construction manager usually provides management services only. It does not self-perform work; all the work is performed by subcontractors. This gives the construction manager the ability to focus on managing and coordinating the work.

The fourth difference is the transparency of the cost of the work. The typical construction management contract provides that the construction manager is paid its general conditions (i.e., overhead) costs and its fee. All subcontractor costs are directly passed through to the owner. The general conditions costs and the fee can be stipulated amounts or they can be calculated as percentages of the cost of performing the work (i.e., percentages applied to the amounts paid to subcontractors).

There are no major systemic disadvantages to the owner in the construction manager at risk method of project delivery. There are, however, a couple of potentially serious operational problems. First, the construction manager at risk may not approach its relationship in a less adversarial manner than a general contractor. There are two reasons for this. The first is that there are a number of companies that historically were general contractors that now describe themselves as construction managers. Their orientation toward their relationships with owners hasn't changed. The second reason is that when construction managers work on projects for owners that only do infrequent construction, they have the same incentives as general contractors to maximize revenues from the project on which they are currently working.

The second operational problem relates to incentives. If the general conditions costs and the construction manager's fee are to be calculated as a percentage of the cost of performing the construction work, then the construction manager has no incentive to hold down costs, and, arguably, has an incentive to let them rise. Similarly, if change orders to the construction manager's contract carry a mark-up for general conditions costs and/or fee based on the amount of the change order, the incentives are not consistent with cost control.

### Fast track

Fast track refers principally to an approach to sequencing the design and construction work in which construction begins before the design is complete. The fast track approach can be part of the design-build approach and part of the construction manager at risk approach. It cannot be used with the design-bid-build approach which contemplates the designs being complete before bidding begins.

The advantage to the owner of using this approach is that it can potentially save time and therefore save money. That is because the total project delivery time is reduced by allowing the design process and the construction process to overlap. There are two potential risks to this approach that make deciding when to use it a more important decision than is sometimes realized.

The first risk is that work that was already completed will have to be partially or completely done over because of the evolution of the design. The changing nature of the design can arise from owner-directed changes or from the designer refining its thinking. The second risk is that in the rush to complete the design documents, or at least to get some portion of them ready for early bidding packages, the documents will be incomplete. This, in turn, will lead to a larger than expected number of contractor change order proposals, many of which may have merit.

Because of these risks, it is prudent to consider using the fast track procurement method in projects that are familiar to the owner, such as standard, repetitive projects. Examples include owners who build gas stations, hotels, or office buildings that are essentially the same building from project to project. In those circumstances, the standardization of the work should minimize the risks inherent in the fast track approach. Owners who build projects infrequently, and public owners with complicated procurement requirements, should be hesitant about using the fast track approach.

If an owner believes it must use the fast track approach because of serious time, cost, and/ or operational constraints, and the project in question is not a standardized, repetitive one for the owner, it should, if at all possible, procure a designer and a contractor that have experience working together. Experience working on a similar type of project is better, and experience working together on a similar type of project for this owner is still better.

The bottom line is that the fast track approach to project delivery places extra management burdens on the designer and the contractor. The owner must be sure its designer and its contractor can meet those burdens.

### *Methods seeking enhanced collaboration*

The owners, designers, and builders of real estate projects have been seeking to reduce ways to reduce conflicts for a number of years. The goal has been to save time and particularly costs that are incurred in resolving disputes. The most costly method of resolving disputes is litigation. The initial effort to reduce costs resulted in a focus on arbitration as a substitute for litigation. As the costs of arbitration rose, the next step was an emphasis on mediation as a way of avoiding the costs of arbitration or litigation. Mediation remains a favorite dispute resolution method for construction projects and is written into many design agreements and construction contracts. Arbitration and mediation represent efforts to streamline the resolution of disputes.

The industry has more recently sought to avoid disputes by encouraging more collaboration from project participants from the beginning of design through construction and occupancy. The first attempt at this approach was partnering. Still in use on many projects, this approach features a facilitated discussion prior to the start of construction at which all the project participants collaborate to establish goals and procedures for the project. Some sessions also jointly discuss project risks. The objective is to align all project participants' goals and to build a collaborative and respectful project approach. The deliverable in this process is a charter which sets forth the agreed upon goals and is signed by all the attendees to signify their commitment. Some projects have a refresher session during the course of construction. On projects where the contractor is brought on board prior to the start of construction, such as construction manager at risk or design-build, there can be an initial partnering session during design and another session prior to the start of construction.

The author is a certified arbitrator and mediator for the American Arbitration Association (currently on inactive status) and has worked on projects where partnering was employed.

Two more recent attempts at developing collaboration are LEAN construction and integrated project delivery, often referred to as IPD. The author has not participated in a project as of this writing where either of the approaches has been used. Therefore, only a very summary description of both is offered.

LEAN construction is focused on maximizing efficiency and minimizing waste in the design and construction of projects. LEAN was originally pioneered in the manufacturing sector and is highly process oriented. It seeks to develop the most efficient way to deliver a design and construction project. The most efficient process has to be determined individually for each project. In order to achieve that objective, there must be a high degree of collaboration among project participants, including the owner, designer, and contractor, and the contractor's key subcontractors. LEAN can be combined with any other project delivery method, but it works best with methods that allow the owner to involve the contractor in the project while design is ongoing. In most cases, this will mean either construction manager at risk or design-build.

Integrated project delivery (IPD) is another approach that seeks to begin project collaboration during design. It requires bringing the contractor into the project early and seeks to align the project participants' goals from the start of the project. The purest form of IPD involves the establishment of a separate entity "owned" by the project owner, the designer, and the contractor. This is to maximize the likelihood that the financial incentives of the three participants are fully aligned. A more common form of IPD is to use a contract which specifies that all the project participants put their fees/profits at risk. If the project finishes on time and on budget, each participant benefits as originally planned. If the project is ahead of schedule and below budget, everyone shares the upside; and, conversely, if the project is behind schedule and over budget, all participants absorb the downside consequences.

In order to achieve the collaboration that both LEAN and IPD seek, these projects, and increasingly all other projects, use a technology known as building information modeling (BIM). While this is not a project delivery method or directly related to contract administration, the reader is encouraged to learn about this technology. BIM provides a digital three dimensional view of the project and its components. As of this writing, the latest BIM is BIM 6. BIM 4 introduced schedule; BIM 5 introduced cost; and BIM 6 is intended to provide project information in a manner that supports building operators and managers so that it provides information and support over the life cycle of the project.

Both LEAN and IPD are quite new. There is a fair amount of literature available on the theory of these approaches, but there are still relatively few projects that have been built using either of them. The reader, if interested, is encouraged to read more about these approaches and to review with a qualified attorney the prospective contractual benefits and challenges.

Several observations are in order. First, any approach that fosters collaboration from an early stage in a project is likely to reduce disputes (more on this thought below). Second, for these types of approaches to work, the owner needs project participants that want to work collaboratively. If you have a designer or a contractor that does not want or know how to work that way, LEAN or IPD terminology, or contract forms won't make that organization collaborative. Third, not every project is a good candidate for either LEAN or IPD. The extra effort required to make these approaches work is easier to justify on large, complicated projects where the owner has the legal flexibility to assemble the team most likely to make the project successful.

*Techniques for enhancing collaboration*

Collaboration between the owner, designer, and contractor is likely to minimize change orders, delays, and disputes. The earlier the collaboration begins, the more likely it is to be successful. The following are techniques that can be employed on any project that will maximize the chances of achieving early and sustained collaboration.

- *The owner must set the tone*. In order to create a culture of collaboration on a project, the owner has to demonstrate its commitment to collaboration by showing it operates in a collaborative manner.
- *Building collaboration should start at the procurement stage*. The requests for proposals for the designer and for the contractor should include a firm's approach to and record on collaboration as an explicit criterion for selection.
- *Insist that meeting agendas include all important issues*. This applies to regular meetings like weekly progress meetings and to special purpose meetings that have a more limited agenda and purpose. Collaboration means working together to arrive at recommendations or decisions. This can't happen on issues that are not included in a meeting's agenda and therefore are not discussed.
- *Ensure that issues are fully explored*. Working together on developing proposed resolution of issues means allowing all parties to have a chance to give relevant input.
- *Insist on collaboration*. If a project participant is not working collaboratively, the owner should be prepared to discuss this with the particular firm. If necessary, there should be a meeting with the firm's senior management to obtain their commitment to fix the problem.
- *Make failure to collaborate costly*. If a firm persists in not working collaboratively, the owner should indicate there will be consequences. Such consequences might include requiring that current project personnel be replaced and/or making clear that failure to improve could influence future opportunities to work for the owner. To demonstrate that there are costs, the owner has to follow through on imposing consequences on a reluctant project participant.

Collaboration is not easy. This is because project participants have different interests (see discussion in Chapter Three). As a result, achieving collaboration requires constant effort. The owner needs to be sensitive to this and be a leader in committing and, when necessary, recommitting to the collaborative approach.

Because project participants have different interests, conducting partnering or calling a project LEAN or IPD won't by itself create collaboration and won't change organizations that don't have a practice of collaboration. That requires a commitment from within each organization.

As a final thought on collaboration, owners should be careful when entering into new project delivery approaches. It is imperative that the owner understand how its interests will be protected under any project delivery method it is considering.

### Project management consultants

The project management consultant approach can be used with any of the other four project delivery approaches discussed in this chapter. The approach involves hiring an individual or a firm to assist the owner to manage the project. While project management consultants were historically individuals known as clerks of the works, this section proceeds on the assumption that today most project management consultants are firms, particularly for projects of any size.

The consultant can be hired at the beginning of the project to assist with all aspects of the project, starting with the selection of the designer, or the consultant can be hired to assist with the management of the construction of the facility. These consultants are known as project managers (usually involved in the entire project), program managers (virtually always involved starting with selection of the designer), construction managers (usually involved just before construction starts), and owner's representatives (involved at either point). The construction manager that acts as a consultant is known as a construction manager as agent, as opposed to the construction manager at risk discussed above.

The advantage for the owner of a hiring a project management consultant is it increases the owner's capacity to manage the project. Owners that don't have in-house project management capacity, or whose capacity is fully utilized while a significant project now needs to get underway can benefit significantly from hiring a project management firm. Typically, the project management consultant is the only participant on the project whose only mission is to protect the owner's interests. In the spirit of full disclosure, for those who didn't read the introduction, the author is a principal in a firm that provides these services and therefore has a predictable bias in favor of such services.

The owner can contract to utilize as extensive or limited a set of services as the owner believes it needs. The owner–project manager contract should make clear, and the owner should insist, that the project manager assembles information and makes recommendations, the owner makes the decisions, and then the project management consultant seeks to make sure the decisions are carried out by the designer and/or the contractor.

The only potential disadvantage to the owner is an operational one. If the project management consultant does not see its role as protecting the owner's interests (because, for example, it is focused on the technical design and construction issues only), it is likely not to add much to the owner's ability to manage the project and the payments to the project management firm will be a waste of the owner's money. Also, the project management firm has a certain incentive to overstaff the job, or sell more services to the owner than the owner needs for that project, because the project manager is also in business to maximize revenues and profit. A project management consultant that appears bent on maximizing revenues in that way should not be hired for, or recommended to others for, subsequent projects.

## Bottom line: owners need same basic protections

Regardless of which project delivery method the owner elects to utilize, the owner needs the same basic protections in its contract or contracts. It needs to be prepared to use the same contract administration techniques. This is because the same basic principle on which this book is based applies regardless of which project delivery method the owner selects. That principle is that the owner and its designer, the owner and its contractor, the owner and its design-builder, and the owner and its construction manager at risk have different organizational and project interests, and therefore protecting the owner's interests through effective contract management is vitally important to the owner.

# 2 Techniques for effective preparation of contracts

## Introduction

One of the best ways to facilitate effective contract management is effective contract preparation. Effective contract preparation means drafting and assembling the contract so as to protect the owner's interests. Protecting the owner's interests when administering a design agreement, a consultant agreement, or a construction contract is a lot easier if these documents are written and assembled with the *intent* of protecting the owner's interests. Therefore, before getting to the specifics of the design agreement, the consultant agreement, and the construction contract, this chapter discusses in general terms techniques for drafting and assembling contracts that protect the owner's interests and provide the fewest possible bases for disputes. Subsequent chapters deal with the substance of design agreements, consultant agreements, and construction contracts; this chapter deals with choice of language and approaches to assembling contract documents.

This chapter discusses ten recommended rules of contract preparation. These rules are listed in Exhibit 2.1.

---

**Exhibit 2.1   Ten rules for preparing contracts**

  1  Develop standard documents.
  2  Create requirements.
  3  Use specific language.
  4  Ask, as drafted, will the owner (and the designer, consultant, or contractor) be readily able to determine if the other party did what the provision requires.
  5  Use language of acknowledgment to establish responsibility.
  6  Eliminate provisions that are duplicative of other provisions.
  7  Assemble the contract with the fewest possible documents.
  8  Incorporate all important documents into the contract.
  9  Do not mix and match between sets of previously prepared contract documents.
10  Create a strong owner position in the contract.

---

## Develop standard documents

The first important step in preparing contract documents that protect the owner's interests is for the owner to develop its own individualized standard documents. These include an owner–designer agreement and an owner–contractor contract. If the owner typically holds certain consultant contracts directly, the standard documents should include an owner–consultant agreement. The standard documents could also include a set of the owner's procedures.

There are three reasons why it is essential for owners who do more than occasional construction to have standard documents. The first reason is to ensure that the design agreement and the construction contract actually protect the owner's interests. If the owner develops these documents, they will protect the owner's control of scope, cost, and schedule; provide appropriate leverage for the owner; and, in general, protect the owner's interests. Documents drafted by designers and contractors, their attorneys, or their trade associations, will not have protecting the owner's interests as a significant objective. Even documents drafted by organizations purporting to represent owners are no substitute for documents drafted by or for an individual owner. Documents drafted for a specific owner will incorporate that owner's preferred approach to managing projects. That is important because it means the owner's preferences are driving project management as opposed to having to adjust one's approach based on the requirements of standard documents prepared by a trade organization.

The second reason it is important to have standard documents is because it will give the owner's project management staff a significant advantage in dealing with designers and contractors. The advantage arises from the in-depth familiarity with the contract documents that comes from using the same documents on project after project. The designer and the contractor will not know the documents nearly as well, even if they perform work for the owner on a regular basis. This advantage provides the basis for analyzing issues in terms of the relevant contract as a complete document, focusing on those provisions that are most relevant to the issue, and understanding how the provisions work best for the owner. This, in turn, facilitates the owner's prevailing on project issues.

The third reason is to ensure that the design agreement and the construction contract relate to each other in all the necessary ways. The provisions of each document that address responsibilities of both the designer and the contractor must require the same things of the designer and the contractor. Examples of provisions that relate to both the designer and the contractor include those dealing with construction inspection, submission and approval of requisitions for payment, and submission and approval of contractor submittals.

## Create requirements

One key to effective contract drafting is creating requirements. That is because it is only possible to establish whether or not the designer or contractor has met a contractual obligation if the obligation is in the form of a requirement. For that reason, it is important that contractual provisions create requirements; not make recommendations or state expectations.

The first step in creating a requirement is to use directive language. This means using words such as "shall," "will," and "must." It also means avoiding words or phrases, such as "should" or "it is expected that," which create recommendations or expectations, but not requirements.

As an example, Exhibit 2.2 shows the three versions of a provision dealing with the designer providing its services according to a schedule included in the design agreement.

The first provision creates a requirement. It states what the designer must do; it must provide its services in compliance with the schedule that is part of the design agreement. The second provision does not create a requirement because it says what the designer "should" do. It "should" provide its services according to the schedule in the design

---

**Exhibit 2.2   Provisions for designer providing services according to a schedule**

1   The Designer shall provide its services in accordance with the schedule included in this Agreement as Exhibit D.
2   The Designer should complete its work according to the schedule included in this Agreement as Exhibit D.
3   It is anticipated that the Designer will provide its services in accordance with the schedule included in this Agreement as Exhibit D.

---

agreement. However, telling the designer what it should do is not the same thing as telling it what it must do—what it is required to do. Similarly, telling the designer what "it is anticipated" will happen—namely, providing its services according to the schedule in the design agreement—is stating an expectation. It is not stating a requirement.

Under the alternative provisions above, what happens if the designer does not provide its services according to the schedule in the design agreement? Under the first provision, the designer is in breach of the agreement. It has failed to perform as required by a contractual provision. Under the second provision, the designer is not in breach of the contract because it has failed only to fulfill a recommendation. Under the third provision, the designer is also not in breach of the agreement because it has failed only to live up to an expectation. That is why creating requirements, rather than recommendations or expectations, is essential to effective contract drafting.

## Use specific language

The most effective way to create a requirement is to use specific language. The more specific a requirement is, the easier it is to determine if the requirement has been satisfied. For example, the period for the contractor to submit a notice of a potential change order might be described in two ways, as shown in Exhibit 2.3

---

**Exhibit 2.3   Provisions on contractor change order notices**

1   The Contractor shall submit a written notice to the owner that it believes it is entitled to additional compensation and/or additional time within a reasonable period.
2   The Contractor shall submit a written notice to the owner for additional compensation and/or additional time within five days from the occurrence of the event or circumstance which the Contractor believes gives rise to such additional compensation and/or additional time.

---

The first provision creates a requirement, but one that is not easy to enforce. The provision says that the contractor must give notice and it must be done within a reasonable period. Because the words "within a reasonable period" are subject to a wide range of interpretation, it would be difficult in many cases to determine if a particular contractual notice complies with the provision. For example, a notice submitted 3 days after the relevant event is probably "within a reasonable period," while a notice submitted 6 months after the event is probably not. Notice submitted any time in between 3 days and 6 months may require subjective interpretation to determine whether it was submitted "within a reasonable period." This will undercut the owner's interest in prompt notice of potential increases in cost as well as its interest in avoiding disputes.

The second provision makes the requirement specific. The period is 5 days, and it starts running from the moment an event or circumstance occurs which the contractor believes entitles it to additional money and/or time. This makes the obligation for prompt submittal of notices clear. It must be submitted within 5 days of the relevant event. It will be easy and straightforward to determine if any notice complies with this requirement. This provision will further the owner's interest in prompt notice and avoiding disputes.

### Ask, as drafted, will the owner (and the designer, consultant, or contractor) be readily able to determine if the other party did what the provision requires

Combining the key point of the first section—make each provision a requirement—with the key point of the second section—make each requirement specific—the owner should review each contractual provision by asking can it be readily determined if the designer or contractor did what the provision requires. From the owner's perspective, a well-written contractual provision is one that allows anyone reading the provision to understand exactly what is required. Consider the provision shown in Exhibit 2.4.

---

**Exhibit 2.4　Provision on contractor schedule**

The Contractor shall submit a schedule which demonstrates how it will complete the project by the contractually specified completion date.

---

This provision does not require any specific type of schedule. Therefore, if the owner wants a critical path method (CPM) schedule, this provision does not provide a basis for requiring one. Indeed, any schedule, no matter how summary, probably complies with this provision. Also, this provision does not specify when the schedule must be submitted. Therefore, it is difficult to determine whether any schedule submission is timely or late.

Depending on what the owner prefers for schedule monitoring purposes, either of the provisions shown in Exhibit 2.5 would be preferable.

---

**Exhibit 2.5　More specific provisions on contractor schedule**

The Contractor shall, within thirty (30) days of award of this Contract, submit to the Owner a bar chart schedule, with a bar for each major trade, demonstrating how the Contractor shall complete the project by the contractually specified completion date.

The Contractor shall, within thirty (30) days of award of this Contract, submit to the Owner a Critical Path Method schedule which shall contain no fewer than one thousand (1,000) activities [this number will depend on project size] and shall demonstrate how the Contractor shall complete the project by the contractually specified completion date.

---

These provisions are specific enough to determine when the schedule is due and whether a schedule submittal meets the technical requirements of the provisions.

## Use language of acknowledgment to establish responsibility

When it is particularly important to document the designer's or the contractor's understanding of a provision and its requirements in order to promote compliance and/or avoid later claims, the appropriate drafting technique is to use language of acknowledgment and agreement. The objective is to make it part of the document that the designer or the contractor understands the requirements of the provision and agrees to comply with its requirements so that it cannot later assert otherwise in order to justify noncompliance or a claim. An example involving the designer might relate to the design schedule, as shown in Exhibit 2.6.

---

**Exhibit 2.6   Provision relating to designer's schedule**

The Designer acknowledges that it has participated in the development of and understands the requirements of the Design Schedule. By executing this Agreement the Designer agrees that it can perform the Scope of Services as set forth in this Agreement in compliance with such Design Schedule.

---

An example involving the contractor might relate to visiting the site prior to executing the contract as shown in Exhibit 2.7.

---

**Exhibit 2.7   Provision related to contractor visiting site**

The Contractor acknowledges that it has visited and carefully examined the Site. By executing this Contract, the Contractor agrees that there are no visible impediments to the completion of the Work on or before the Completion Date set forth in the Contract.

---

The importance of this type of provision is its ability to preclude claims based on a designer or contractor asserting that it did not understand the requirements of a provision, or that it did not believe the provision contained a material requirement of the agreement. However, this type of language should be used sparingly. Frequent use will undercut the owner's ability to rely on the acknowledgment and agreement language. If the language appears in a lot of provisions, it will not be viewed as a thoughtful, intentional statement by the parties related to a particular provision or a few provisions. It will be seen as standard contract language and of no particular effect.

## Eliminate provisions that are duplicative of other provisions

It is not unusual for different parts of the contract documents to have provisions that address the same subject. Frequently, these provisions don't say precisely the same thing. This creates ambiguity and/or inconsistencies, either of which can cause problems for the owner. For example, the general conditions prepared by the owner might have a provision

---

**Exhibit 2.8 Provisions on contractor quality control**

Provision in General Conditions:

- The Contractor shall, prior to submitting its first requisition for payment, submit to the Owner for the Owner's approval a plan describing how the Contractor shall implement quality control on this project. The Plan shall include, among other things, the employment of a full-time quality control manager, the use of a Nonconforming Work Report ("NWR") which shall be prepared by the Contractor, shall describe work that does not comply with the Contract Documents, and shall be issued to the relevant Subcontractor, the Designer, and the Owner. Each NWR shall be considered closed only when signed off by the Designer and the Owner.

Provision in General Requirements (Division 1 of the Specifications):

- The Contractor shall be responsible for implementing the Contractor's quality control program and for ensuring the quality of the Work.

---

addressing quality control, while the general requirements prepared by the designer might have another provision on quality control as illustrated in Exhibit 2.8.

The contractor may argue that it is not required to provide a full-time quality control manager or to submit a plan and use NWRs because of the provision in the general requirements. The contractor will argue that the extent of its responsibility for ensuring quality is set forth in the provision in Division 1. That responsibility is to implement its own quality control program with no input from the owner.

The best solution to this type of problem is to remove provisions that are duplicative. There are no issues addressed by the contract documents that need to be dealt with twice. Each requirement should be described once in the contract documents. There are two approaches that will make achieving this goal more likely. The first is to combine the owner–contractor agreement and the general conditions into one document. This document can be called the owner–contractor contract. The second is to limit the general requirements section of the specifications (usually, Division 1) to those matters not covered in the owner–contractor contract.

To further mitigate this type of problem, it is advisable to include an order of precedence provision in the contract. An order of precedence provision indicates the order of priority that the provisions in the various contract documents will have in the event of conflict. If the owner–contractor agreement has the highest priority, followed by the general conditions, the specifications, and then the drawings, then the contractor will have to provide a plan and use NWRs in the above example. If the order of precedence provision gives higher priority to the specifications than to the general conditions, then the contractor will not have to provide a plan for the owner's approval or use the specified form.

### Assemble the contract with the fewest possible documents

Another technique for minimizing ambiguities and inconsistencies is to create contract documents that include the fewest possible separate documents. This is particularly relevant for short form contracts with the designer, a consultant, or the contractor. These short form contracts are often used for smaller projects where the owner and the designer, consultant, or contractor believe it is not necessary to have a long, involved contract.

When short form contracts are used, it is not unusual to describe the scope by attaching a proposal submitted by the designer, the consultant, or the contractor. Sometimes the business terms are also described by attaching a proposal. The problem arises from the potential inconsistencies between the owner's standard short form contract, and the terms of the proposal.

The proposal may describe the scope of services as including more services or construction than the owner wants. The proposal may describe essentially what the owner wants, but the owner may intend something slightly, but importantly, different.

Another problem is that proposals frequently have disclaimers that are not in the owner's interests. For example, the owner's standard contract may contain language requiring the invoice or requisition for payment to be in a certain format or include certain documentation to substantiate the amount claimed. The proposal may not include that language. The owner's contract may provide for the withholding of retainage; the proposal may not. The proposal may limit the liability of the entity submitting the proposal. The owner's contract will probably not contain such a limitation.

To avoid these types of problems, it is often best for the owner to prepare its short form contract and its own attachment using as much, or as little, of the proposal's scope of services or scope of work language as it wants. That way the potential ambiguities and inconsistencies between the owner's short form contract and the designer's, consultant's, or contractor's proposal can be avoided.

The potentially most problematic version of this problem is when the designer, consultant, or contractor attaches its standard business terms to its proposal. The owner should detach those immediately and inform the proposer that the agreement will be documented by use of the owner's contract form. In the great majority of cases, the other party will not object. If they do object, that should be a strong warning signal to the owner. That is because resistance to the owner's contract form implies that the proposer has terms that are significantly different than are in the owner's document. The owner should focus on those differences to determine if they are deal breakers, subject to negotiation, or more apparent than real. If they are deal breakers, then the owner should insist on using its contract form even at risk of not completing the procurement with the preferred designer or contractor. If the differences are, in the owner's judgment, subject to negotiation then the negotiation should lead to acceptable modifications to the owner's contract form, not use of the designer's, the consultant's, or the contractor's form. If the differences are more apparent than real, the result should be the same, acceptable modifications to the owner's contract.

The owner should always insist on using its contract form, rather than agreeing to modifications in the designer's, the consultant's, or the contractor's form. That is because the owner has a contract carefully drafted to protect its interests. It is used to administering its own contract. These are significant advantages for the owner. These advantages are not significantly diminished by negotiating acceptable modifications to the standard form. The advantages are likely to be significantly reduced, or even eliminated, by use of the designer's, the consultant's, or the contractors, standard form.

## Incorporate all important documents into the contract

Some contracts treat certain documents as reference documents. These documents are referred to, but not incorporated in, the contract. There are two categories of reference documents. The first are documents that contain information that may be of interest to the designer or contractor. An example of such a document in the design agreement is a project program prepared by another designer or a consultant. An example in the construction contract is a geotechnical report containing information about underground site conditions. The second type of reference document is one that contains technical information that relates to owner preferences. An example of such a document in either the design agreement or the construction contract is an owner manual on the procurement and maintenance of telecommunications systems.

Treating a document as a reference document potentially creates an ambiguity as to the extent of the designer's or contractor's responsibility. The first type of reference document contains information. For such a document, the question becomes whether the designer or contractor is contractually responsible for the information in the referenced document. For example, if the construction contract references a geotechnical report, is the contractor responsible for knowing whatever information is in the report? To put the question another way, can the contractor rely on the information in the report, and, if so, can the contractor use the difference between the information in the report and actual below ground conditions at the site as the basis of a differing site condition claim?

These ambiguities are best resolved by explicitly incorporating in the contract any document that contains information for which the owner wants the designer or the contractor to be responsible. For information that the owner wants the designer or contractor to know but for which the owner does not want to be responsible, it is advisable to reference the document together with a strong disclaimer. Exhibit 2.9 illustrates this approach for a geotechnical report.

---

**Exhibit 2.9   Provision referencing document with disclaimer**

The Designer, through its geotechnical consultant Soils Engineering, has conducted an investigation of below ground conditions at the Site. The results of such investigation are available upon request in the form of a report entitled "Conditions at the 1200 Main Street Site" dated October 17, 2015. The data contained in the report is provided only for the Contractor's information, and the appropriate use of such data, if any, shall be determined by the Contractor in its sole discretion. The Owner does not warrant the accuracy of the data nor any conclusion which may be stated in such report or inferred from such data.

---

The second type of reference document adds to the technical specifications. It contains information related to the designer's or the contractor's required performance. The construction contract might include a reference to the owner's manual for its telecommunications system. The contract might include the manual in a section entitled "Other documents." This treatment makes ambiguous the extent to which the contractor is responsible for the requirements in the manual. The manual might contain such requirements as system standards, the types of equipment to be used, the acceptable brands of those types of equipment, and methods of installation of equipment to be installed in the owner's building. If the owner wants the contractor to perform its work in compliance with the manual, there are two approaches that are preferable to simply referencing the document. The first, and the most prudent, is to have the firm that prepares the electrical or low voltage specification lift the requirements from the manual and incorporate them word for word in the appropriate

---

**Exhibit 2.10   Provision incorporating a document into the contract**

The [Owner's Name] Manual for Telecommunications Systems dated April 23, 2016 ("The Manual") is hereby incorporated in the Contract Documents. The Contractor, by submitting its bid, represents that it has read and understands all applicable requirements of the Manual.

---

specification. The other alternative is to explicitly incorporate the manual in the contract by a provision like the one in Exhibit 2.10.

## Do not mix and match between sets of previously prepared contract documents

In the interests of creating quality documents, and in the interests of efficiency in the document preparation process, owners sometimes create contract documents by using all of, or portions of, preprinted contracts. This includes such things as:

- using, as is, a design agreement and a construction contract authored by different trade associations;
- using, as is, a design agreement and a construction contract prepared by different entities;
- assembling the design agreement and/or the construction contract by using parts of two or more preprinted contracts; or
- assembling the design agreement and/or the construction contract by combining owner-drafted provisions with provisions taken from one or more preprinted contracts.

Preparing agreements in this manner creates a significant risk of creating ambiguities and inconsistencies in the contract documents. The best way for the owner to mitigate the potential ambiguities and inconsistencies is to draft its own set of contract documents and to use them consistently whenever it undertakes the design and construction of a project. If the owner's staff reads another contract document that it thinks contains certain provisions that are better drafted than those in its current contract, it should integrate the provisions of that document into its own contract documents carefully by studying the effect each new provision has on the entire set of contract documents.

The owner should resist the temptation to mix and match in any of the ways described above. Taking portions of documents, or an entire document out of a set of documents, and marrying them to a portion of the owner's set of contract documents is virtually certain to create ambiguities and/or inconsistencies. Any apparent gain in drafting quality or efficiency is highly likely to be lost to potential reduced protection of the owner's interests as well as increased likelihood of performance period disputes.

## Create a strong owner position in the contract

When drafting the design agreement and the construction contract, there is always a need to balance several considerations. These include, among others:

- the owner's need to control project scope, cost, and schedule;
- the need to allow the designer and the contractor sufficient latitude to use their skill, expertise, and judgment to give the owner the project it wants;
- the need for quality designers and contractors to view the project as one on which they can make a reasonable profit; and
- the need to achieve the owner's project-specific objectives.

The drafting challenge is how to balance these and other considerations. No matter how the owner elects to balance the various relevant considerations, it is important that the owner's contract documents create a strong contractual position for the owner. There are two reasons for this recommendation. First, the owner needs to maintain control of the project's scope, cost, and schedule. This will not happen simply because the owner is the owner. The owner needs certain contractual provisions in order to maintain control.

The second reason for creating a strong contractual position is to ensure protection for the owner if the relationship with the designer or contractor becomes problematic. If problems arise with the designer or the contractor, the owner will need contractual provisions that give the owner tools with which to solve the problems. If, for example, the designer designs a project in excess of the owner's budget at the designer's initiative, the owner will need to be able to require the designer to revise the design documents to bring the project within budget and to do so at no extra cost to the owner. Similarly, if the contractor falls behind schedule for reasons that are the contractor's responsibility, the owner will need to be able to direct the contractor to recover schedule using whatever means are necessary at no additional cost to the owner. These are just two examples of problems that may arise with the designer or the contractor for which the owner needs the contractual authority to resolve the problem in its own best interests.

As a business decision in the interests of the project, the owner may elect to relinquish one or more of its rights under the design agreement or the construction contract. That decision might be made for at least two reasons. First, the owner may perceive that the relationship with the designer or contractor is a productive one. If the relationship is going well, the owner might reasonably decide not to insist on every right it has under the contract in the interests of not negatively impacting a productive relationship. For example, if the designer's invoices are not accompanied by the documentation required by the design agreement, but the owner believes the invoices reasonably reflect the work being done, and the design work is proceeding on schedule and on budget, then the owner might not make the designer submit all the missing documentation.

The second reason the owner might not insist on every right it has under the contract is to provide an incentive to the designer or the contractor to meet the owner's budget or schedule when it appears that one or the other may be missed. For example, if the contractor is somewhat behind schedule for reasons that are not the owner's responsibility, and the schedule is vitally important to the owner, the owner may offer to relinquish its right to assess contractually authorized liquidated damages if the contractor can complete the job within two weeks of the contractual completion date.

These are business decisions that the owner may make for good reasons. They are not positions forced on the owner for lack of contractual authority. That brings us to the key point here. The owner can relinquish its rights under the design agreement or the construction contract whenever it believes it is in its interests to do so. It cannot acquire rights that are not in the design agreement or the construction contract when the contract is executed. That is why it is important to build a strong owner position in these contracts. It is possible to relinquish rights one has; it is not possible to acquire rights one does not have.

# 3  Techniques of contract administration

## Introduction

This chapter introduces contract administration. It discusses the objectives of administering design, consultant, and construction contracts from the owner's perspective and explains why contract administration is so important to the owner. The chapter then discusses a range of techniques for contract administration. The intent is to increase the owner's effectiveness at contract administration. When reading this chapter, the reader should keep several things in mind. First, these techniques apply equally to administering the design agreement, the consultant agreement, and the construction contract. While the scopes are different, the suggested techniques apply equally to each of these contracts. Second, the objective of contract administration is the successful completion of the owner's project. These techniques are intended to support that objective, not to become ends in themselves. Third, the use of each technique should be carefully tailored to the circumstances of each project. The techniques should not be used in a "cookie cutter" fashion on every project. A productive relationship with a designer, consultant, or contractor should not be over managed. Conversely, an owner faced with a difficult designer, consultant, or contractor should use all of the applicable techniques to proactively protect its interests.

## Objectives of contract administration

The objective of contract administration for the owner is to obtain all the benefits of the bargains it has made with its designer and its contractor. To amplify this thought a little bit, it is useful to step back and look at the traditional model of contract formation. A contract requires an offer and an acceptance. To be binding, the contract must contain mutual consideration; that is, there must be value for both parties. For each party to a contract, the value offered by the other party is the benefit of the bargain.

   The benefits to the owner occur at different levels. Before beginning the design and construction of a facility, the typical owner will undertake a planning process for that facility. This process will establish the scope, budget, and schedule parameters. The benefits to the owner from its design agreement and construction contract are the successful completion of a project within the planned-for level of scope, budget, and schedule.

   Each of these agreements contains specific benefits to the owner. The benefits provided by the design agreement include the plans and specifications to be prepared by the designer. However, the design agreement typically includes additional benefits, such as inspection of the work, review of shop drawings and other contractor submittals, review of change orders, and convening and chairing weekly progress meetings. Similarly, the contract with the contractor describes specific benefits to the owner. The principal benefit is the completed construction of the facility described by the plans and specifications. The benefits are also likely to include providing quality control during construction, providing insurance or cooperating with the owner's controlled insurance program, and, particularly for public owners, compliance with minority and women participation goals at the workforce and/or subcontractor level.

The objective of contract administration is to ensure that the owner obtains all the benefits of the bargains it has made with its designer, its consultant(s), and its contractor. There are a number of other aspects of project management involved in obtaining these benefits, including overall project management, construction inspection, estimating, scheduling, and others. However, contract administration is an integral part of the owner's effort to ensure it obtains all the benefits for which it has contracted.

## The heart of contract administration: translating benefits to deliverables

If the objective of contract administration is to make sure that the owner receives all the benefits to which it is entitled, then the heart of contract administration is identifying and obtaining those benefits. In order to obtain these benefits, the owner must make them tangible. The process of contract administration moves prospective benefits off the pages of a contract into the physical and organizational realities of the project.

The first step is to identify the benefits. It is necessary to read the design agreement, the consultant agreement, and the construction contract to determine what responsibilities each agreement requires the designer, the consultant, and the contractor to meet. It is these specific responsibilities established by the design agreement, the consultant agreement, and the construction contract that translate the owner's benefits into the specific deliverables to which it is entitled. In general, these deliverables will fall into five categories. These categories are listed in Exhibit 3.1.

---

**Exhibit 3.1   Categories of contract deliverables**

The deliverables of design agreements, consultant agreements, and construction contracts relate to:

1   performance of technical requirements;
2   project administration;
3   how services will be performed;
4   the business relationship; and
5   issue and dispute resolution.

---

The first category of deliverables relates to the performance of technical requirements. For the designer, this category relates principally to services related to preparing plans and specifications. For the contractor, this category involves construction of the facility as described by the plans and specifications.

The second category of deliverables is services related to project administration. For the designer, the principal service in this category is construction administration. The deliverables are the services involved in effectively administering the construction on the owner's behalf. These include such activities as inspecting the work to make sure it complies with the contract documents, reviewing requisitions for payment, approving submittals, and evaluating change orders. For the contractor, this category relates to what are commonly called submittals. There are three groups of submittals. The first are administrative submittals relating to management of the project. An example is the submission of a schedule of all submissions required of the contractor. The second group is construction submittals, principally shop drawings, and product samples. The third group is project closeout submittals. Examples include warranties, and operation and maintenance manuals.

The third category of deliverables consists of requirements that relate to how services will be performed. For example, the design agreement might include a schedule for the completion of the project documents. For the contractor this category includes such things as quality control and safety plans. In some projects, it might include compliance with a project labor agreement.

The fourth category of deliverables involves requirements related to the business relationship between the owner and the designer and the contractor. For example, these agreements typically specify, or authorize the owner to establish, the format and documentation required in order for requests for payment to be processed and paid. The deliverables are the invoices and requisitions for payment that comply with these requirements.

The fifth category of deliverables relates to how issues and, when necessary, disputes will be resolved. The most frequent issue is requests for additional compensation. The design agreement typically specifies how the designer is to request compensation for additional services. These provisions address the format, substance, and documentation required for these requests. Similarly, the construction contract describes the process for requesting additional compensation and/or extensions of time, including the format, substance, and documentation required for these requests to be addressed. These agreements also describe the process for resolving issues that cannot be successfully negotiated. These deliverables provide the owner an orderly process for addressing issues, the ability to control the process of resolving the issues, and the ability to control the outcome of the process, subject only to arbitration or litigation.

Having identified the deliverables, the second step in translating benefits to deliverables is obtaining them. In many cases, the designer and contractor will submit the deliverables as required at their own instigation. Sometimes the owner will have to request the deliverables. Often such requests will be sufficient to obtain the deliverables. Sometimes, simple requests will not be enough; then, the owner will have to approach obtaining deliverables as a process of enforcing contractual requirements.

After the owner receives the deliverables, it must determine if they comply with the applicable contractual requirements. Even the most sophisticated owners will need assistance to determine if certain technical submittals meet contractual requirements. For example, whether proposed penetrations of structural steel and concrete for mechanical pipes or electrical conduits require additional support or reinforcing of the structural elements is an issue to be resolved by an expert. On the other hand, most owners can readily determine whether the documentation accompanying a request for payment or a request for additional services or a change order proposal is sufficient to substantiate requested amounts. The point in either case is that the owner is entitled to a submission that complies with contractual requirements. A key step in the contract administration process is to determine that deliverables actually meet the applicable requirements of the contract.

The deliverables that the owner is entitled to under the design agreement and the construction contract are all intended to work together to achieve for the owner its objectives related to scope, cost, and schedule. The owner needs effective contract administration to protect its objectives related to scope, cost, and schedule. These objectives govern the benefits for which the owner contracted. The benefits take the form of specific deliverables in the contract documents. The owner must identify these deliverables and make sure it receives the deliverables, and receives them in the way required by the contract.

## Designer and contractor have different interests

In the context of actual projects, the designer and the contractor frequently wish to provide these benefits in a different manner than that anticipated by the owner. Even though the design agreement and the construction contract specify the benefits and the deliverables in what is supposed to be the manner agreed to by the parties, the designer and contractor may attempt to modify, reduce, or even eliminate those benefits during the performance of their contracts. That is because the owner and its designer, and the owner and its contractor, have different interests. It is why effective contract administration is so critical to owners and has such an important impact on whether or not a project is successfully completed.

It is important to bear in mind while reading this section that these differences in interests exist between professional and ethical organizations. As the differences are described below, the point is not that the designer and/or the contractor have different professional or ethical standards than the owner. To the contrary, these differences in interests exist despite the professionalism and common ethical standards of all project participants. Of course, if the designer or the contractor act unprofessionally or unethically, the divergence in interests will be even greater.

The owner and the designer have a common interest: they want to complete the facility. The owner wants the facility to use for its intended purpose. The designer wants to complete the project to be able to use it in the designer's marketing material and to get paid for the work.

The owner and the designer are likely to have different interests with respect to scope, cost, and schedule. The owner has determined a level of quality, functionality, and aesthetics it needs for its facility. The designer may believe that a higher level of quality and/or aesthetics would make the project more successful. This difference can impact not only scope, but the cost and the schedule of both the design and construction work.

The designer has different interests than the owner with respect to the cost and schedule of its own work. A designer with a lump sum agreement will have an incentive to do the least possible work consistent with its contractual responsibilities. Conversely, a designer with a cost plus arrangement has an incentive to maximize its level of effort. The owner's interest in both cases may be the opposite of the designer's. In the lump sum situation, the owner may seek (and be entitled to) a greater level of effort; while in the cost plus situation, the owner may want (and only need) a more limited level of effort.

The designer has no reason—independent of its contractual responsibility—to favor the same schedule as the owner. In most cases, the owner is interested in starting and completing its project as promptly as possible. The designer may want to proceed more slowly for such reasons as cash flow management (e.g., current cash flow is acceptable, but future cash flow is more problematic), or because it proposed a staff member or subconsultant that is an expert on some aspect of the project design but is not yet available to work on the project.

The owner and the contractor also share an interest in completing the project; in the owner's case, so the facility can be used for its intended purpose, and, in the contractor's case, so that it can get paid and can use the project in its marketing efforts. In many cases, the owner and the contractor also have a common interest with respect to schedule. That is because in the typical lump sum situation, the sooner the contractor completes the project, the greater its profit is likely to be.

The owner and the contractor have different interests with respect to the contractor's scope of work and compensation. In the context of a lump sum agreement, the contractor will want to limit the quality and extent of the work. If the contract, or a change order, is priced on a cost plus basis (or time and materials in the case of a change order), the contractor has an incentive to maximize level of effort. The owner's interests are likely to be the opposite. The owner will seek (and may be entitled to) greater quality and a fuller extent of work. Conversely, with respect to cost plus (or time and materials) work, the owner may not want (and may not need) the highest level of effort.

It is these differences in interests that make the contract administration function so critical to the successful completion of a project from the owner's perspective. If all the project participants had the same interests, then contract administration would need to be little more than an administrative function, moving paper and/or electronic transmissions to the appropriate entities. However, because of these differences in interests, contract administration, from the owner's point of view, is a key function in protecting the owner's interests and conserving the owner's resources.

Having discussed contract administration in general terms, the chapter will now address specific techniques of contract administration. Twelve techniques are outlined below which will be explained in more detail in the rest of this chapter.

1  *Reading the contract* Since all of the owner's rights, responsibilities, and, most important, benefits, are based on the design agreement and the construction contract, it is vitally important to carefully read these documents prior to the start of any project.
2  *Identifying and obtaining benefits* The owner needs to know how to identify and actually obtain the benefits included in its contracts. This means obtaining benefits when the designer and the contractor are cooperating, and when one or both are resisting.
3  *Preparing effective correspondence* Properly structuring correspondence can materially assist the owner to resolve issues in its best interests.
4  *Interpreting the contract* There are important rules of contract interpretation that are applicable to resolving issues with the designer and the contractor. There are also techniques of contract interpretation that the owner can use to its advantage.
5  *Understanding and utilizing contractual leverage* The design agreement and the construction contract typically provide the owner with important points of leverage. An owner that understands and effectively uses this leverage can greatly enhance its ability to resolve issues in its best interests.
6  *Understanding and avoiding waivers* What the parties do, what they write, and what they say during the course of the performance of a contract term can alter the terms of the contract. It is important that the owner understand how these waivers occur and how to avoid them.
7  *Assembling and managing documentation* Complete documentation is critical to successfully resolving contractual issues. The owner must understand the role of documentation, which documentation is particularly important, and how to manage documentation.
8  *Developing procedures* For owners that undertake construction projects on a regular basis, or for an owner that is undertaking a very large project, the management of the

project is greatly assisted by having explicit procedures that establish how important processes on the project will take place.

9   *Using meetings effectively* Most construction projects, including projects in the design phase, involve a virtually endless number of meetings. The owner can increase the likelihood of achieving its objectives, as well as saving time and money, by understanding effective meeting management.

10   *Identifying and winning designer and contractor games* Some designers and contractors use certain strategies which I have characterized as games. It is important to be able to recognize these strategies and to defeat them.

11   *Not pointing an unloaded gun and other miscellaneous techniques* There are a few additional techniques which can assist the owner to effectively administer its contracts.

12   *Minimizing disputes* The owner can reduce the number of disputes involved in a project by following three rules.

## Reading the contract is the first step

The first step in administering a contract is to read it very carefully. This is not meant as a facetious comment. Many people charged with administering contracts do not take the time to read the document in its entirety at the beginning of the job. That is a mistake.

Reading the contract should involve more than reading the titles of provisions or skimming the text. The contract should be read carefully, provision by provision. The reader should determine the meaning of each provision and should consider how the provision relates to other provisions. For example, how do the clauses describing what constitutes additional services for the designer relate to the description of basic scope; or how does the provision requiring the contractor to keep working relate to the provision allowing the contractor to dispute a denial of a change order proposal.

There are at least four reasons to read the contract thoroughly at the beginning of the project. First, it is important to know the contract in detail. The ability to recognize the owner's rights in any given situation depends on a thorough understanding of the contract. Second, for those owner personnel that work for an organization that has standard construction documents and are themselves experienced at contract administration, it is still wise to read the contract beginning each new job. That is because the organization's legal, purchasing, or other department responsible for promulgating the documents may have changed them. Third, reading the document before starting the project is a good way of making a habit of going to the contract to get answers to issues where there is a potential or actual disagreement with the designer or with the contractor. Fourth, constantly reading the contract will place the owner contract administrator in a strong position vis à vis the designer or contractor. That is because the owner staff member will know the document much better than the designer and the contractor and will be able to more effectively use it to the owner's advantage.

## Identifying and obtaining contract deliverables

The process of identifying and obtaining contract deliverables generally involves three steps, and often includes a fourth step. The first step is to read the contract. The second step is to identify the deliverables. The third step is to request the deliverables, if they have not been submitted as required, and the fourth step is to obtain the deliverables in circumstances where the designer or the contractor fails to submit the required deliverable notwithstanding repeated requests for it.

Every item that the designer and the contractor are required to provide to the owner is a deliverable. That term is commonly used in connection with design contracts. While every item that the contractor is required to provide the owner is also a deliverable, it is more usual to refer to these items as submittals. For ease of use and to avoid repetition, we will refer to all such items under both types of contracts as deliverables.

The second step in administering a design or construction contract is to identify the deliverables the designer or contractor is required to provide to the owner. The best way to do this is to develop the list while reading the contract, as described above. Each provision should be analyzed to determine if it requires one or more deliverables. Some provisions are easy to analyze. For example, the provisions that require the designer to provide schematic designs, design development documents, and construction documents all specify required deliverables. But, so do the provisions that require the designer to assist the owner to obtain permits, to attend progress meetings, and to prepare minutes for those meetings. Similarly, the technical specifications that require the contractor to furnish and erect certain specified structural steel, drywall, and pavement clearly set forth deliverables. But, so do the provisions that require the contractor to prepare and implement a quality control plan, prepare and update a schedule, and provide certain types of insurance.

For the reader that is not familiar with this process, any provision that requires the contractor to submit something to the owner is a provision that establishes a requirement for a deliverable. Typical wording of this type might be something like the following.

> Contractor shall submit to the owner, for the owner's approval, a plan setting forth how Contractor shall perform the quality control activities required by this Paragraph.

Some of the most important deliverables that contractors are usually required to submit include the following:

* schedule of submissions;
* schedule of values;
* baseline schedule and progress updates;
* quality assurance/quality control plan;
* safety plan; and
* shop and coordination drawings.

This approach should result in a comprehensive list of deliverables. Exhibit 3.2 is an example of a comprehensive list of deliverables. This chart was developed and utilized on a $550 million project where the project delivery method was construction manager at risk.

Having identified the contractually required deliverables, the owner must now take the necessary steps to make sure it receives the deliverables. The first task is to turn the list of deliverables into a format that allows the owner to track if each required deliverable has been submitted. Exhibit 3.3 illustrates such an expanded format.

Unfortunately, life on a construction project is never simple; designers and contractors frequently do not submit required deliverables within the contractually specified time. In that case, the next step for the owner is to request the missing deliverable. The manner in which the owner makes its request should depend on the designer or contractor's response. There are three levels of making such a request.

The first level is informal with no record. This is typically a phone call to the person on the designer's or contractor's team with immediate responsibility for producing the

## Exhibit 3.2    Construction manager's deliverables

| Document | Paragraph | Deliverable | Due date (if applicable) |
|---|---|---|---|
| CM agreement | 5.3 | Written reviews of drawings, specifications | |
| | 5.5.2 | Update construction cost estimates | At least 3 estimates between 5/1/99 and 12/31/99 and at least 4 between 1/1/00 and 12/21/00 |
| | 5.6 | Schedule of procurement of long-lead time items | |
| | 5.8 | Prepare permit application materials as required by Exhibit PER | |
| | 5.10 | Monthly Progress Report, including: executive summary; project status overview; photographs; procurement status report; project schedule update; project cost update; data required by Section 5(1) of Chapter 152; status report on CM's project staff; and list of outstanding issues | On or about 15th of the month for the preceding month |
| | 5.11.2 | Procure subcontracts according to this paragraph when prequalification is not required | |
| | 5.11.3 | Copies of all executed subcontracts | Promptly |
| | 5.11.6 | Prepare bid tabulations | |
| | 5.11.6 | Schedule of values for each subcontract | Promptly upon execution of subcontract |
| | 5.13 | Detailed organization chart | Within 30 days of execution of CM Agreement |
| | 5.14 | Full time quality control manager | |
| | 6.5.4 | Separate accounting system for change orders | |
| | 7.4 | Notice of the availability of cash discounts | |
| | 10.2 | Certificate of completion of work in accordance with contract | Upon completion of the work |
| | 10.13.1 | Projected cash flow schedule | Within 120 days of execution of CM Agreement |
| General conditions | 3.2.2 | Full time site superintendent with a builder's license | |
| | 3.2.2 | Name, telephone and beeper numbers of a CM representative available 24/7 and never more than an hour from the site | |
| | 3.2.9 | Mechanical, electrical, plumbing, and sprinkler coordination drawings | |
| | 3.2.14 | Construction Management Plan | 2/15/00 |
| | 3.3.8 | Certified payrolls of CM and subcontractors | Weekly |

| Document | Paragraph | Deliverable | Due date (if applicable) |
|---|---|---|---|
| | 3.4.3 | Written warranty of all CM work; same warranty from each sub | Prior to substantial completion |
| | 3.4.6 | Names, telephone numbers for emergency service on MEP equipment | Prior to start of warranty period |
| | 3.8.1 | Originals of permits and other approvals | Upon completion of the work |
| | 3.8.3 | Record drawings for the entire work | Prior to final payment |
| | 3.9.2 | Schedule of submittals for each bid package including a schedule for shop drawings and a schedule for samples | |
| | 3.9.12 | Shop drawing log | |
| | 3.14.1 | Copies of all correspondence to the designer | |
| | 3.17.5 | Surveys establishing locations of improvements and utilities | |
| | 3.18.1 | Progress photographs (20 plus 4 aerials) | Monthly with Monthly Report |
| | 3.19.1 | Operations and maintenance manuals (4 copies) | Prior to final payment |
| | 5.1.3 | Project directory with names, addresses, and telephone numbers of key subcontractor personnel | |
| | 7.2.1 | Draft CPM construction schedule | Within 90 days of execution of CM agreement |
| | 7.2.1 | Bar chart summary schedules; work plan (looking ahead schedules) | Monthly |
| | 7.2.1 | Initial baseline CPM schedule | Within 30 days of receiving comments on the draft CPM schedule |
| | 7.2.4 | Revise cost loaded baseline schedule as contracts are awarded | Within 60 days of award of subcontract |
| | 7.2.5 | Progress schedule update (of baseline CPM schedule) | Monthly (with requisition for payment) |
| | 7.2.5 | Narrative report on schedule compliance, update of the bar chart summary schedule, and work plan for next month | Monthly (as part of monthly report) |
| | 8.1.1 | Schedule of values | Before first payment under CM Agreement |
| | 8.1.1 | Updates to schedule of values | Upon award of each subcontract and prior to requiring any payment for that subcontractor |
| | 8.9.1 | Punch list | When CM thinks work is substantially complete |
| | 8.10.5 | Full project documentation | Prior to final payment |
| | 8.11.2 | Inventory of material stored off site | Monthly |
| | 9.1 | Full time safety officer | |
| | 9.2.1 | Site security program | Prior to notice to proceed |
| | 9.2.10 | Suitable and adequate fire protection equipment | |

| Document | Paragraph | Deliverable | Due date (if applicable) |
|---|---|---|---|
| | 9.3.1 | Certificate that CM did not bring/release any hazardous materials on site | Upon completion of the work |
| | 10.3.6 | Certificate of insurance for policies required by the CM Contract | |
| | 11.1.5 | List of pending change orders and unresolved claims | Monthly (as part of monthly report) |

**Exhibit 3.3   Tracking form for contractor deliverables**

| Document | Paragraph | Deliverable | Due date | Follow up action (date) | Received (date) |
|---|---|---|---|---|---|
| CM agreement | 5.13 | Detailed organization chart | Within 30 days of execution of the contract | Email requesting overdue organization chart (3/3/03) | |
| | | | | Letter requesting overdue organization chart (3/24/03) | |
| | | | | Second letter requesting overdue organization chart (4/9/03) | Received (4/16/03) |

deliverable. This is, in most instances, the appropriate place to start because it is non-confrontational and promotes a productive relationship with the designer or contractor. In a relationship which is basically working well, this rationale suggests making at least two informal requests before moving to the next level of request.

The second level of request is informal with a record. This might be an item at a regularly scheduled progress meeting. It might be the subject of a meeting. In either case, the request for the deliverable should be included in the meeting minutes together with a reminder of the designer's or contractor's obligation to produce the deliverable. The item in the meeting minutes might read as follows.

[Name of the owner organization] stated that [name of the contractor] has not yet submitted its safety plan, as required by the Contract. It was agreed that [name of contractor] would submit the plan in two weeks [preferably, a specific date].

Another form of making an informal request with a record is to send the responsible person, the person who was initially called, an email requesting the deliverable. The email would be informal in tone, but would be printed out to create a record of the request. An example of such an email might be as follows.

Susan:

    As we discussed a couple of times in the last couple of weeks, [name of contractor] was supposed to have submitted a safety plan by now. Please get it in to me by [specific date]. We need to review this plan and get everyone's sign off because it is required by the OCIP [i.e., owner's controlled insurance program] on this project.

    Thanks for your help.

Frank

The third level of request is formal with a record. This is usually a letter. Exhibit 3.4 contains an example of such a letter.

---

**Exhibit 3.4   Letter requesting overdue deliverable**

Widget Manufacturing Company
125 South Street
Anytown, State 01112

July 14, 2016

XYZ Construction Corp.
10 Main Street
Anytown, State 01111

Attn: Ms. Susan Smith; Project Manager

**Re: Widget Manufacturing Company / Construction of New Facility / Safety Plan**

Dear Susan:

This is in reference to the safety plan that XYZ Construction Corp. ("XYZ") is contractually required to provide to Widget Manufacturing Company ("Widget").

    This requirement has been discussed at each of the first five progress meetings (June 4, June 11, June 18, June 25, and July 2, 2016). On each of these occasions, XYZ stated that it would be submitting the plan in the next day or two. In addition, on June 27, Widget emailed a request to XYZ. On the same day, XYZ responded that it would submit the plan later that day. To date, XYZ has failed to submit a safety plan. XYZ is requested to submit the required plan within the next three business days.

Sincerely,

Frank Jones
Project Manager

---

The Preparing effective communications section contains detailed recommendations on how to structure effective correspondence. The point here is to illustrate how the owner can formalize its request for a deliverable, and, if necessary, enforce the contractual requirement to submit the deliverable. That brings us to the next stage, which is enforcing contractual requirements.

    When a designer or contractor fails to comply with a contractual requirement, the owner has two choices. It can seek to enforce the requirement, or it can make a business decision to accept noncompliance. A third alternative is to continue to request the submittal but not use the enforcement approach described below. Except in unusual circumstances, the owner should avoid this third alternative. Pursuing this approach makes the owner appear at best indecisive, and at worst weak, and will encourage further lack of cooperation from the designer or contractor.

There are various business reasons for not insisting that the designer or the contractor satisfy a contractual requirement. The three most common are changed circumstances, preserving a productive relationship, and costs significantly exceeding benefits.

The way a project develops, or the stage the project is in, may render a contractual requirement less important than anticipated when the agreement was drafted. For example, the designer may be obligated to submit a cost estimate at each stage of design. This requirement may become less important at the construction document stage if the owner elects to start construction before the construction documents are completed, or if the owner engages a contractor before the design documents are completed, and the contractor's estimate confirms that the project can be built within the owner's budget. Similarly, the contractor's obligation to submit a monthly update of its project schedule may become less important in the later stages of the project, if the project is clearly on schedule.

If the relationship with the designer or the contractor is generally a productive one with which the owner is satisfied, the owner may wish to avoid upsetting this relationship. Under these circumstances, it may elect not to enforce a contractual requirement, particularly if it's not critical to successful completion of the project.

The third reason for electing not to enforce a contractual requirement is that the costs of enforcement may outweigh the benefits. For example, the contractor is often obligated to counter sign daily time sheets submitted by any subcontractor to document time and material charges. If the contractor submits the sheets signed by a subcontractor but not countersigned by the contractor for work the owner agrees was performed on the days claimed, the owner may conclude that forcing the contractor's superintendent to review these sheets in detail would be a distraction from progressing the project with no substantive benefit to the owner.

Returning now to instances where the designer or contractor refuses to produce contractually required deliverables, and the owner wishes to enforce the requirement, it is important for the owner to understand how contractual requirements are most effectively enforced. The next step (following those described above and specifically following the letter request illustrated in Exhibit 3.4) is to inform the other party of the consequences of continued refusal. Returning to the example of a contractually required safety plan, if the efforts described above have not produced submission of a plan, the next step is to describe the potential consequences of continued refusal to submit a safety plan. Exhibit 3.5 illustrates how this letter might be drafted.

There are several important points worth noting about the letter in Exhibit 3.5. First, it enumerates the history of the owner's efforts to obtain the safety plan from the contractor. As will be addressed in the Preparing effective communications section, when confronted with an uncooperative designer or contractor, it is important to enumerate the history of the owner's efforts to obtain the deliverable to which it is entitled. That history will help tell the owner's story if the issue develops into a full fledged dispute. Second, this letter now tells the contractor explicitly what the consequences are of not submitting the required deliverable, and states the owner's intention to impose those consequences.

Third, note the statement is that the contractor has not submitted the submittal; as opposed to stating that the owner has not received the document. There are two reasons why it is preferable to state that the contractor has not submitted something rather than that the owner

**Exhibit 3.5   Letter requesting overdue deliverable stating consequences**

Widget Manufacturing Company
125 South Street
Anytown, State 01112

July 28, 2016

XYZ Construction Corp.
10 Main Street
Anytown, State 01111

Attn: Ms. Susan Smith; Project Manager

**RE: Construction of Riverside Facility / Safety Plan**

Dear Susan:

This is in reference to the safety plan that XYZ Construction Corp. ("XYZ") is contractually required to provide to Widget Manufacturing Company ("Widget").

This requirement has been discussed at each of the eight progress meetings held to date (June 4, June 11, June 18, June 25, July 2, July 9, July 16, and July 23, 2016). On each of these occasions, XYZ stated that it would be submitting the plan in the next day or two. In addition, on June 27, Widget emailed a request to XYZ. On the same day, XYZ responded that it would submit the plan later that day. On July 14, Widget sent XYZ a letter requesting submission of the safety plan. To date, XYZ has failed to submit a safety plan.

Paragraph 8.5 of the General Conditions states in part:

> Contractor shall prepare and submit to Owner for Owner's approval a plan stating in reasonable detail the Contractor's approach to managing safety on this project. Such plan shall be submitted and approved prior to the submission of Contractor's first requisition for payment.

XYZ's safety plan is overdue. As has been discussed at length at the various progress meetings, because of the height and unique shape of this facility, this project has serious potential safety issues. Therefore, until XYZ has submitted and Widget has approved a safety plan, Widget will be forced, pursuant to the provisions of Paragraph 8.5, to reject XYZ's first (and any subsequent) requisition for payment.

Sincerely,

Frank Jones
Project Manager

has not received it. The first is that the preferred formulation states that contractor has failed to do something it is required to do, as opposed to something simply not happening. The second reason is that stating that the owner has not received something leaves open the possibility that the problem lies with the delivery system, rather than with the contractor, a less descriptive position.

If the contractor continues to resist satisfying the requirement, the next step for the owner is to impose the stated consequences. The imposition of the consequences should be fully documented in follow up correspondence.

## Preparing effective correspondence

The legal and business relationships on a project are between the owner and its designer, and the owner and its contractor. They are not between individual project participants. The owner's project manager and the designer's project manager, for example, do not have a legal or business relationship. They may, and hopefully do, have a productive relationship based on shared goals and mutual respect, but, if either has to leave the project, the legal and business relationship between the owner and the designer continues. More importantly, the performance and financial responsibilities are assumed by the owner and the designer, not by any member of either's staff. A commitment to perform a scope of work, or to pay for the performance of that scope, is a commitment undertaken by the owner or the designer as an entity, not a commitment undertaken by either project manager individually.

Project correspondence should reflect this reality. Letters, memoranda, or any other type of correspondence intended to be official should be from entity to entity. A letter heading should contain an attention line with the name and title of a person to whom the letter is addressed, as well as a regarding line (i.e., "RE: . . .") stating the subject of the letter. The subject of the letter would typically include the name of the project and the specific subject of the letter. Exhibit 3.6 illustrates the suggested format.

---

**Exhibit 3.6    Appropriate letter heading**

May 6, 2016

Sunrise Construction Co., Inc.
245 Elm Street
Anytown, State 01112

Attn: Ralph Jackson; Project Manager

**RE: Anytown High School / Contractor's Required Baseline Schedule**

Dear Ralph:

---

Because it is the owner, designer, and contractor, as business and legal entities, which have the responsibility to perform commitments, correspondence should be written in the third person, not the second person. When the owner sends a letter to Sunrise, attention Ralph Jackson, responding to a letter dated June 13, written by Jackson, the owner's response should refer to "Sunrise's letter of June 13," not "your letter of June 13." Similarly, if the owner's letter refers to something that Jackson, or Burt Wilson, the assistant project manager, said at a meeting, the reference should be to "Sunrise's commitment to submit a baseline schedule the next day," not to "your commitment" or "Burt's commitment." If it becomes necessary to use the name of the individual, the person's title in the designer's or contractor's organization should follow the name (e.g., Ralph Jackson, Sunrise Construction's Project Manager). This is to make clear that the person referred to made a statement or took an action in the person's official capacity as an agent of the designer or contractor.

The selection of the persons to receive copies of the communication is also important. Sending copies of communications is an effective way to advise all those parties that need to be informed of project developments. In the case of a dispute, carefully selecting recipients will be important as a way to put all appropriate parties on notice of the owner's position. Since most owner organizations are political (in the sense that how people interact, or perceive that they are interacting with others is important to them), it is also wise to carefully consider who in the owner's organization would expect to get a copy of the correspondence, and/or who would be annoyed or slighted if they did not receive a copy.

An effective communication is one that persuasively states the owner's position. How to construct such a communication will be discussed in detail and an example provided. First, however, it is important to place the preparation of correspondence in the context of the resolution of disputes.

Disputes that are not resolved at the project level or by senior management through negotiation will be resolved by a third party. A third party, for these purposes, is a person (or panel of persons) authorized by contract and/or by law to impose a resolution of the dispute. The two most common third persons are judges and arbitrators. Government projects also involve third parties with titles such as hearing officers and boards of contract appeal. There are two important points about third persons. First is that they are authorized to resolve the dispute (subject in most cases to some level of legal appeal). Second, they will resolve the dispute many weeks or months, often years, after the events giving rise to the dispute have taken place. For that reason, most third parties will rely heavily on the available documentation.

Because the third party can, and will, resolve the dispute, and because the third party will rely extensively on documentation, it is vitally important that the owner's documentation be as persuasive as possible. The documentation will need to influence the view of someone who does not have first hand knowledge of the events. This means that when drafting a piece of correspondence that relates to an issue that is, or may become, in dispute, it is very important to think about how it would be interpreted by a third person. Because it is important to influence the third party, and because filing systems, logistics, and other aspects of case management are imperfect, each piece of correspondence should stand alone. That is, it should be carefully constructed in the manner described below.

A well-constructed piece of correspondence has four components: a statement of purpose, a discussion of factual matters, an analysis of relevant contractual provisions, and a conclusion. Each will be discussed in turn and an example provided, based on a circumstance where the contractor has not provided a contractually required schedule to the owner, Quick Shop Grocery Stores, in connection with the construction of a supermarket.

The statement of purpose describes the purpose of this letter, and why it is being written. Exhibit 3.7 illustrates a statement of purpose.

---

**Exhibit 3.7   Statement of purpose**

This letter is in reference to the obligation of Sunrise Construction, Inc. ("Sunrise") to submit a construction project schedule for review and approval by Quick Shop Grocery Stores, Inc. ("Quick Shop"). Specifically, this letter responds to Sunrise's Baseline Project Schedule Rev 2 dated June 27, 2016.

The next component of the communication is a discussion of the facts. This discussion should have two parts: the factual history of the issue to date and the current factual aspects of the issue. The factual history describes what has taken place with respect to the issue to date. In many cases, this will be a recitation of the owner's efforts to obtain a specified deliverable and the contractor's lack of, or insufficient, response. The current factual aspects generally deal with the details of why the latest action (or lack of action) by the designer or contractor is not acceptable. This approach is illustrated in Exhibit 3.8.

---

**Exhibit 3.8   Discussion of the facts**

On April 17, 2016, Sunrise submitted a proposed Baseline Project Schedule. On April 25, Quick Shop rejected this initial proposed Baseline Project Schedule because it contained too few activities. On May 5, Sunrise submitted Baseline Project Schedule, Rev 1. On May 15, Quick Shop rejected Baseline Project Schedule, Rev 1 because, although additional activities were added as required, the durations for the mechanical, electrical, and plumbing ("MEP") trades were unrealistically short. On June 2, a meeting was held at Sunrise's request to discuss the contract scheduling requirements and how Sunrise could best meet them. On June 16, Quick Shop called Sunrise to request that it submit Baseline Project Schedule, Rev 2 within the next five days. On June 27, Sunrise submitted Baseline Project Schedule, Rev 2.

The MEP durations, although extended from Rev 1, remain unrealistically short. Completion by the required completion date as shown on Rev 2 is in doubt because of the unrealistic MEP durations. Furthermore, Rev 2 shows painting beginning in certain areas where the drywall patching and taping will not be completed; this is also unrealistic.

---

These two paragraphs illustrate the two parts of the factual discussion. The first paragraph describes the history of the Baseline Project Schedule submission. The second describes the current facts involved in reaching a conclusion on the issue. This is the suggested format for all pieces of correspondence: the factual part of the letter should discuss historical facts and current facts.

In cases where the designer or the contractor is unresponsive on an issue, it is important to describe this history in detail. The description of this lack of cooperation will paint a picture for a third party, if one becomes involved, and will set a context for consideration of the issue by the third party.

When circumstances require the owner to write a series of letters on the same subject, it becomes tempting to skip the history. First, it saves time to avoid creating the history; and, second, it can appear "in your face" and unnecessarily antagonistic to keep repeating the facts showing the designer's or contractor's failure to respond as required. The temptation to skip the history should be avoided. Every piece of correspondence prepared on an issue that is, or may become, in dispute should stand on its own. Every piece must be as persuasive as possible for the third person involved in resolving a dispute. Therefore, it is important to repeat the history, and expand it as necessary to be current, in each piece of correspondence. The picture painted, and the context set, for the third party becomes clearer with each letter, if the history is properly described. Because of the "cut and paste" function in computer word processing programs, repeating the history is not as burdensome as it used to be.

After the facts, come the relevant contractual provisions. The relationship between the owner and designer is governed by the design agreement. Similarly, the relationship between the owner and the contractor is governed by the construction contract. Therefore, any position taken by the owner should be based on the relevant contract. Exhibit 3.9 illustrates the appropriate discussion of relevant contractual issues.

---

**Exhibit 3.9    Discussion of contractual provisions**

Section 01310 of the Specifications requires that Sunrise submit a Baseline Project Schedule for review and approval by Quick Shop. Paragraph 1.03(A) of that Section requires, in part, that the Baseline Project Schedule "contain realistic durations for all activities," and Paragraph 1.03(B) requires that the Schedule's logic be reasonable "and that specific predecessor–successor relationships support the overall schedule logic."

---

When discussing contractual provisions, two approaches are effective. If a provision is helpful, the letter should cite it and summarize what it says. If it is really helpful, the letter should quote it, or quote the pertinent part that is particularly supportive of the position that the letter is taking. What should not be done, is writing that "the contract requires" with no citation to one or more specific provisions. That approach is not persuasive and will not be viewed favorably by a third person.

The final part of a well-structured piece of correspondence is the conclusion. This is where the owner establishes its position on the issue. The conclusion should be clearly stated and tied to the facts and analysis set forth in the previous parts of the letter. Exhibit 3.10 illustrates a statement of the owner's position.

---

**Exhibit 3.10    Statement of owner's position**

Baseline Project Schedule Rev 2 does not comply with the contractual requirements for Sunrise's baseline schedule. The durations of MEP activities remain unrealistically short. The logic of certain painting activities is not supportable. Therefore, Baseline Project Schedule Rev 2 is rejected. Sunrise is directed to submit Baseline Project Rev 3 within 15 days.

---

Exhibit 3.11 illustrates a complete letter written in the recommended manner. This letter deals with an important issue, particularly on projects built in urban neighborhoods: site security.

## Interpreting the contract

This section deals with legal rules of contract interpretation and with techniques for interpreting the contract in the manner most advantageous to the owner. The rules are established by law and apply to everyone; the techniques are recommended methods of analysis and interpretation.

There are a number of rules of contract interpretation that have been developed over the years. In fact, the United States inherited its common law legal system from Great Britain,

**Exhibit 3.11 Complete letter on site security**

Major City Bank
1 Bank Plaza
Major City, State 04128

March 16, 2016

Atlantic Construction Management Co., Inc.
23 Day Boulevard
Major City, State 04128

Attn: Mr. David Wilson; Project Manager

**RE: Major City Bank Office Building / Site Security**

Dear Dave:

This is in reference to Atlantic Construction Management's ("Atlantic") contractual obligation to provide site security.

On February 14, 2016, Major City Bank ("Major City") observed the gate open at approximately 8:00 PM and secured and locked the gate. The problem was communicated to Atlantic the next day. For some period of time thereafter, Atlantic locked the gate at least by 8:00 PM.

On Monday, February 26, Wednesday, March 7, and Tuesday, March 13, Major City observed the site gate on Elm Street open at approximately 5:45 PM, but no work being performed and no Atlantic or subcontractor personnel on the site. On Thursday, March 15, Major City observed the gate unlocked (the lock appeared to be missing) at approximately 8:00 PM. This means that the site is not secured. On each of these dates, Major City sent an email to Atlantic noting the lack of security and requesting corrective action.

Now that the site contains a number of pieces of construction equipment, it is very likely to become a place where children will want to play, but where there is a foreseeable danger that they will be injured as a result of playing on the site. The site may attract curious adults, putting them at risk of injury as well.

The lack of total site security is inconsistent with the provisions of the Construction Manager's Security Program ("the Program") submitted by Atlantic on August 3, 2000. Paragraph 2.1(a) of The Program states that Atlantic will "insure a complete enclosure." Paragraph 2.1(h) states:

> As the perimeter fencing is one of the key elements of the security program, the condition of the fencing will be reviewed on a regular basis to check for evidence of entry or signs of deterioration due to normal wear and tear or construction activity.

Paragraph 9.2.1 of the General Conditions requires Atlantic to provide security services.

> Without limitation, the CM shall provide security watch service at all such times as are necessary to protect the interests of the CM and the Owner and to provide

for the safety and security of the general public, employees and agents of the Owner and the Designer, and other persons who may be affected by the Work, and to exclude unauthorized persons from the site.

In previous discussions of Project security needs, Atlantic has taken the position that security guards are not necessary at this stage of the construction, and that fencing and locking the site is sufficient for the time being. Major City has repeatedly expressed its intent that the site be adequately protected. First, Major City is concerned for the safety of neighborhood residents and visitors. Second, a neighborhood resident seriously injuring him/herself on the site could have serious negative consequences for the Project, particularly if it were perceived that the Project had not taken adequate steps to prevent this foreseeable problem.

Major City believes it is time to resolve the extent and timing of security services. Therefore, Atlantic is requested to submit a revised plan within the next five (5) days for providing effective security for the site. If Atlantic believes that guard services are still unnecessary, it should state when and to what extent guard services will be provided and explain how the site will be secured (including who will be responsible for locking the gates) until then so that no unauthorized entries can be made. Until this issue is satisfactorily resolved, Major City will not approve any further payments to Atlantic for security.

Sincerely,
Major City Bank

James MacDonald
Vice President—Capital Facilities

cc: William Davidson, MCB
Mary Howard, MCB
Fred Wellington, Designer

so these rules have been around for a long time. The purpose in describing the rules here is to assist owner personnel in understanding how a third person would engage in contract interpretation.

Exhibit 3.12 states the five rules that have recurring applicability to design and construction contracts.

---

**Exhibit 3.12   Five rules of contract interpretation**

- The contract will be interpreted to mean what it says.
- The contract will be interpreted as an entire document.
- Only if there is ambiguity as to the meaning of a contract term, will matters outside the contract be relevant.
- Provisions prepared specifically for the contract in question will be given precedence over standard terms.
- Ambiguities will be interpreted against the drafter of the contract.

Most court decisions dealing with contract interpretation will start with the idea that the contract should be interpreted to give effect to the intent of the parties. The next question becomes how to determine the parties' intent, particularly when the parties are differing over what that intent was. The rules summarized in Exhibit 3.12 are used, then, to arrive at the parties' intent.

The first rule is that the contract will be interpreted to mean what it says. Words will be given their plain meaning, unless given a specific meaning by the contract. For example, the word "work" used in a construction contract will mean the activities performed by the contractor to build the facility required by the contract. The term "Work," however, will mean whatever the contract defines the term to mean. Such a definition might be the labor, materials, equipment, supervision, and all other resources necessary to meet the requirements of the specifications and drawings and to meet all other requirements of the contract documents.

The difference in the two meanings concerns the extent to which the definition of work ties the contractor to the nontechnical requirements of the contract. For example, the requirements to provide a safety plan and a quality control plan would typically be in the general conditions portion of the contract documents. So the word work will not make providing these two plans part of the contractor's work, whereas the specifically defined term "Work" will include the requirement to provide these two plans.

The second rule is that the contract is to be read as an entire document. This means that all the terms of the contract must be read to determine the resolution of an issue. For example, in one contract with which the author is familiar, the term "subcontractor" was defined in one provision as any entity that had a contract with the contractor, or with another subcontractor, to perform a portion of the work. Another provision defined subcontractors for purposes of certain payment rights as only first tier subcontractors with subcontracts having a value in excess of $500,000. So, the term subcontractor had one meaning for payment purposes and a more inclusive meaning for all other purposes.

The third rule of contract interpretation is that only if there is ambiguity as to the meaning of a contract term will matters outside the contract be relevant. This rule is particularly important because many designers and contractors will want to use industry practice as the basis for determining what a contractual provision means. If, for example, the agreement calls for payment of designer invoices 45 days after receipt and acceptance by the owner, the owner has 45 days to make payment, even if the usual practice in the area is 30 days. If, on the other hand, the agreement says that the owner will pay designer invoices "within a reasonable time following receipt and acceptance by the owner," then, if a dispute arises as to when payment should be made, it would be appropriate for the designer to show what industry practice is in the area of the project.

The fourth rule is that provisions prepared specifically for the contract in question will be given precedence over standard terms. This rule applies when the parties take a standard contract and attach additional provisions. This is frequently done when the American Institute of Architects (AIA) documents are used because attorneys representing both owners and contractors want to protect their client's interests by adding certain additional provisions. Often the additional provisions are contained in a document called "Supplementary Conditions" or "Special Conditions." When a provision in the Supplementary Conditions is inconsistent with a provision in the standard document, the provision in the Supplementary Conditions is given precedence because it was drafted specifically for the project in question. The rule operates to give the provision in the Supplementary Conditions precedence even if the provision in the Supplementary Conditions was written five years ago by the relevant attorney and is inserted in every agreement that the attorney reviews. That is because it is considered as being specific to the project as compared to an entire standard document.

Another circumstance in which the rule in favor of provisions drafted specifically for the particular project prevails occurs when a custom written Division One is inserted in the technical specifications of a construction contract. Division One provisions typically deal with project management issues. If a provision in the Division One specifications is inconsistent with a provision of the general conditions, and the general conditions are a standard document, and the Division One specifications are custom written for the project, the provision in the Division One will prevail.

The fifth rule of contract interpretation is that ambiguities will be interpreted against the drafter of the contract. If a provision in a contract is subject to two reasonable interpretations, the interpretation offered by the party who did not draft the contract will prevail. That is important for owners who prepare, as this book strongly recommends, a standard set of contract documents because it will mean that ambiguities will be interpreted in the designer or contractor's favor. That means that it is very important to ensure that the owner's documents are clear and unambiguous.

The above rules are legal rules that have been established by courts. The next few paragraphs offer several techniques that assist the owner to interpret contracts effectively. These techniques are listed in Exhibit 3.13.

---

### Exhibit 3.13   Techniques of contract interpretation

- Understand the owner's preferred position.
- Don't assume the other party is correct.
- Read the entire contract.
- Read provisions carefully to understand what they really say.
- Defined terms are capitalized.
- Consider reasonable and unreasonable interpretations.

---

The first technique is to understand the owner's preferred position. Contract interpretation should be approached as an exercise in advocacy. The owner has a preferred position in virtually every situation in which the owner and the designer, or the owner and the contractor, have recourse to the contract documents to determine how to resolve an issue. It is important to start by understanding what that position is.

Having understood the position, the rest of contract interpretation is really a process of determining how to support that position most effectively using the contract documents. Occasionally, reviewing the contract documents will disclose that they unambiguously support the designer's or the contractor's position. In those rare cases, the owner should acknowledge that and resolve the issue as favorably as possible. However, and this is the second technique, the owner should not assume the other party is correct. No matter how confident they appear, or how smoothly they quote the contract documents, don't assume they are correct. It has been the author's often-repeated experience that parties who take positions with apparent conviction are wrong, or have a position that is weaker contractually than the owner's position.

The third technique is to read the entire contract. In many instances, there will be more than one provision that relates to the issue being considered. It is necessary to read them all. The fourth technique is related: read each of the provisions carefully and make sure to understand what they say. Often, contractual wording is complicated, and the first reading will not produce the precisely correct understanding of what a provision actually says. At least the provisions that relate directly to the issue in question should be read several times to be sure they are correctly understood.

In determining the meaning of contractual provisions, it should be remembered that one of the legal rules described above is that words will be given their common sense or usual meaning, unless the contract assigns a word a specific meaning. In most cases, words defined in a contract are capitalized. The fifth technique is to identify applicable defined terms. They are readily identifiable. For example, if the design agreement refers to "additional services," a reasonable interpretation is that what is being referred to are services not in the original scope. However, if the reference is to "Additional Services," then the reader knows that the term is defined in the agreement, and that the specific definition will determine whether services performed by the designer are part of the base scope or not.

The sixth technique is to consider reasonable and unreasonable interpretations of relevant provisions. One starts by assuming the owner's position is reasonable and then determines if the relevant provisions can be interpreted reasonably to support that position. Then one explains why other plausible interpretations of the provision are unreasonable or less reasonable. Exhibit 3.14 illustrates this technique using a memo that addresses a design agreement's requirements relating to errors and omissions insurance and dispute resolution.

---

**Exhibit 3.14    Memo arguing interpretation of provisions of design agreement**

**MAJOR CITY BANK**
**Memorandum**

TO: James MacDonald, Vice President
FROM: Mary Howard, Assistant General Counsel
DATE: January 16, 2016
CC: William Davidson, Project Manager
SUBJECT: Renewal of Project Professional Liability Policy; Commencement of Legal Action

This follows up on our conversation of this morning in which we addressed two issues:

* if Major City Bank does not renew the Project Professional Liability Policy, what happens; and can we start suit before finishing mediation if we think there are statute of limitation problems?
* what happens if we can't or don't renew the Liability Policy?

On the one hand, it appears Major City Bank is obligated to provide this policy for the duration of the project. On the other hand, the Agreement can be interpreted as saying, whether we do or not, the Designer is still liable to Major City Bank for any of its errors or omissions.

Paragraph 10.1.2 states (in part):

The Bank has procured and will maintain a project professional liability (errors and omissions) policy (the "Project E&O Policy") for the Project.

---

This appears to obligate us to provide the Policy for the entire duration of the project. Paragraph 10.1.2 also states:

> The Designer's practice professional liability policy (the requirements for which are described in Subsection 10.1.1) shall be maintained in full force and effect with no exclusion of the Project from coverage thereunder, and the limits of such practice policy shall apply in excess of the limits of the Project E&O Policy.

This provision (as well as Paragraph 10.1.1) appears not to provide for the replacement of the Project E&O Policy with the Designer's policy if we don't renew our policy. Paragraph 10.1.2 further states:

> The Bank's provision of the Project E&O Policy shall in no way relieve or limit, or be construed to relieve or limit, the Designer or any subconsultants of any responsibility, obligation or liability whatsoever otherwise imposed by this Agreement or arising out of performance of the services.

This provision can be interpreted as saying, whether or not there is insurance in place, the Designer is fully liable to the Bank for its mistakes. This language is not crystal clear, however. It refers to providing the Policy as not limiting liability. It would have been better for us if the provision had said, "Neither the Bank's provision of the Project E&O Policy, nor the Bank's failure to provide such a Policy shall relieve or limit . . . ." However, the opposite construction appears to make less sense. The opposite interpretation would require us to read this provision as saying, in effect, "If the Bank fails to provide a Project E&O Policy for any portion of the Project, the Designer shall be relieved of its liability [or relieved of its liability to the extent of the amount of the missing Project E&O Policy] for errors or omissions occurring during that time period." That might have been a reasonable proposition if worded directly, but it seems like a real stretch from these words, which appear to have the opposite intent.

Our preferred interpretation of this language might give the Designer the incentive it needs to agree to a modification in the insurance language to insure coverage.

Can we start suit now?

The answer appears to be, "Yes." Again, the answer is not crystal clear.

Paragraph 8.3 states (in part):

> Unless otherwise agreed by the Bank and the Designer, all claims, disputes and other matters in question between the parties to this Agreement arising out of or relating to this Agreement or the alleged breach thereof which are not resolved pursuant to the process described in Sections 8.1 and 8.2 shall be submitted for resolution to a court of competent jurisdiction in Large County, Large State. No such action shall be brought, however, until the completion of all services under this Agreement or the earlier termination hereof as provided in Article 9, unless the continued deferral of filing such action would result in such claim, dispute or other matter in question being barred by statutes of limitation or repose.

This provision can be read to allow filing to avoid statute of limitations problems as applying only to the requirement that legal action not commence before the Designer is done with its work. This interpretation would require that we complete the dispute resolution process of Article 8 before commencing suit, regardless of statute of limitation problems. The other, better interpretation is that the provision allows filing to avoid statute of limitations problems regardless of the status of the dispute resolution process. This is the better interpretation because the second sentence (i.e., "No such action . . .") appears to provide two, separate qualifications to the preceding sentence [which requires a special meeting, then mediation, then legal action]. The first qualification is that, even if we finish the dispute resolution process as described, we have to wait to bring suit till the Designer is finished with its work. The second is that we don't have to wait to bring suit if we have a statute of limitation problem.

## Understanding and utilizing contractual leverage

The design agreement and the construction contract contain various terms that give leverage to the owner. It is very important that owner personnel understand the basic concept of leverage, be able to identify where in the contract the leverage points are, and feel comfortable using that leverage in appropriate circumstances. Each of these aspects will be discussed.

Leverage, for these purposes, means the owner is contractually authorized to withhold something of value, or take other steps with negative economic impact, to the designer or contractor if the designer or contractor fails to comply with the relevant contractual requirements. The most typical form of leverage is to withhold payment for all or a portion of an invoice or requisition for payment.

The key to identifying provisions that create leverage is to look for provisions that require owner approval, place a burden on the designer or contractor, or explicitly allow the owner to withhold payment. Exhibit 3.15 illustrates several provisions that create leverage.

---

**Exhibit 3.15    Examples of provisions that create leverage**

The Contractor shall submit a quality assurance/quality control plan for the review and approval of the Owner. Such review and approval shall be completed by the Owner prior to the submission of the Contractor's initial request for payment.

The Designer shall submit, for the Owner's review and approval, a list of permits required on the Project. Such review and approval shall be completed prior to the Designer's submission of its second invoice, provided that the first invoice shall be restricted to Mobilization Costs.

[As part of longer provision defining Additional Services] The Designer shall have the burden of showing it is entitled to such Additional Services and that the amount it seeks for such services is the amount actually owed to it by the Owner.

---

The most typical points of leverage in a design agreement are in the provision authorizing the owner to withhold payment. This provision usually authorizes the owner to withhold payment for failure to perform design services as required by the design agreement. It may authorize withholding for additional reasons, such as failure to pay subconsultants. The authority to withhold payment for failure to perform services as required by the design agreement is actually quite broad. It may, depending of course on how it is worded, apply to the entire scope of the designer's services. Even if the provision is qualified to apply to "material failure" to perform the services as required, it still authorizes withholding in a wide variety of circumstances. It probably applies to at least the following circumstances:

- failure to provide contractually required design deliverables;
- failure to provide design deliverables according to a schedule included in the design agreement;
- providing design documents (plans and/or specifications) that are incomplete, inconsistent, and/or erroneous (e.g., don't comply with applicable codes); and
- failure to provide assistance in permitting or other project related activities as required by the design agreement.

The owner should be willing to use this leverage in its dealings with a designer where the relationship is difficult. For those interested in the theoretical justification for using the leverage, it goes back to the basics of contract formation. A binding contract results when one party makes an offer, the second party accepts the offer, and there is consideration (something of value) promised by both parties to the other party. The promise by both parties to provide something of value to the other is also frequently referred to as an exchange of promises. When one party fails to deliver what it has promised, the other party is also relieved of its relevant promises. Reduced to its simplest concept, the design agreement represents a promise by the designer to provide design services to the owner and a promise by the owner to pay the designer for those services. So, if the designer fails to provide some portion of the services it promised to provide, the owner is legally and morally justified in not paying for it.

To be optimally effective, leverage must be used carefully. There are several reasons for this. First, the designer must have actually failed to meet the contractual requirement, as opposed to looking like it might fail, or worrying out loud that it might not be able to meet the requirement. Second, using leverage may upset an otherwise productive relationship. Third, the use of leverage will not be legally supportable unless the designer has actually failed to meet the requirement. Failing to use leverage with a designer that is not meeting its obligations is to leave the owner substantially at that designer's mercy.

The first step in the use of leverage should be a formal notice to the contractor of the intention to use leverage. The second step is to actually follow through and use the leverage. The third step is to communicate formally to the designer that the action has been taken and what the designer must do to receive the withheld payment (or to rescind whatever other action the owner may have taken). Exhibit 3.5 illustrates the first step. Exhibit 3.16 below illustrates the third step (which can only occur if the second step is actually taken).

Using leverage with the contractor involves the same approach. It begins with identifying the points of leverage. These tend to be more numerous in a construction contract. Often a construction contract will authorize withholding payment when the contractor fails to submit:

- an acceptable schedule of values;
- an acceptable baseline schedule or schedule update;
- documentation as required to support a requisition for payment;

- documentation as required to substantiate a change order proposal; and
- safety and quality control plans required by the contract.

The owner, under most construction contracts, has a more indirect but very significant form of leverage in connection with contractor claims. The contract usually says that the burden of substantiating a claim lies with the party making the claim, or the contract more directly says the burden of substantiating a claim lies with the contractor. Either way, the contractor is contractually assigned the burden of proving its claim. This means that the owner does not need to disprove a claim, and does not need to pay the contractor on account of any claim until the contractor has satisfied the burden of proving entitlement and value.

When the owner intends to exercise a form of leverage under the contract, it should precede the exercise with at least one written warning as illustrated by Exhibit 3.5. There are two reasons for such a warning. The first is fairness; the contractor should be informed of the consequences of continued failure to comply with a requirement. The second reason is that the warning should serve as an incentive for the contractor to comply with the requirement, and compliance at that point is a lot more efficient than having to continue pursuing the issue. If the contractor does not respond to the warning, then the owner, having given the warning, should exercise the leverage and take the step it said it would. Then, having taken the step, the owner should inform the contractor of that action and indicate what the contractor must do to reverse the owner's action. Exhibit 3.16 illustrates such a communication.

---

**Exhibit 3.16   Letter following withholding of payment**

Widget Manufacturing Company
125 South Street
Anytown, State 01112

August 11, 2016

XYZ Construction Corp.
10 Main Street
Anytown, State 01111

Attn: Ms. Susan Smith; Project Manager

**RE: Construction of Riverside Facility / Safety Plan**

Dear Susan:

This is in reference to the safety plan that XYZ Construction Corp. ("XYZ") is contractually required to provide to Widget Manufacturing Company ("Widget").

This requirement has been discussed at each of the eight progress meetings held to date (June 4, June 11, June 18, June 25, July 2, July 9, July 16, and July 23, 2016). On each of these occasions, XYZ stated that it would be submitting the plan in the next day or two. In addition, on June 27, Widget emailed a request to XYZ. On the same day, XYZ responded that it would submit the plan later that day. On July 14, Widget sent XYZ a letter requesting submission of the safety plan. Again on July 28, Widget sent XYZ a letter requesting submission of the safety plan and indicating that XYZ's first progress payment would not be processed until receipt of XYZ's safety plan.

Notwithstanding all of the discussions and correspondence described above, XYZ has still not submitted its safety plan. Therefore, because of the importance of the safety issue, and XYZ's safety plan, pursuant to the provisions of Paragraph 8.5 of the General Conditions, Widget declines to accept XYZ's first requisition for payment, which was received yesterday. This requisition will not be processed until XYZ has submitted its safety plan, and it has been approved by Widget.

Sincerely,

Frank Jones
Project Manager

A less drastic approach is to withhold partial payment. The best way to facilitate partial withholding is to require the contractor's schedule of values to itemize activities of particular importance to the owner (such as, in the example, the safety plan) and to assign a monetary value to each such activity. Then, if the contractor fails to perform an activity (i.e., fails to submit a deliverable), the owner can withhold the amount specified in the schedule of values.

## Understanding and avoiding waivers

The owner can easily change contract requirements during the performance of the contract by what the owner does, writes, and/or says. This is known as waiving a contractual requirement. While, in theory, the law requires a waiver of a contractual requirement to be knowing and intentional; in practice, courts will readily find intent, particularly in circumstances in which the owner took action, or failed to take action, and the designer or the contractor relied on that act or failure to act. This issue of waiver is hugely important to owners. Waivers occur on projects every day, and many of them are not in the owner's best interests.

The owner can waive contractual requirements by what it does (or fails to do), what it writes, and what it says. This applies to all parts of the contract documents, from the agreement and general conditions to the technical specifications. Examples include the following.

- *What the owner does* The most common form of this type of waiver is the approval, or failure to reject, a contractor's proposal to use a different approach to performing the construction work. For example, if the contract requires that electric panels be connected to the power source with 500 Kcmil cables, and the contractor proposes using 600 Kcmil cables (because this size is safer and/or less prone to system problems), if the owner agrees to accept the 600 Kcmil cables, it has waived the contractual requirement for 500 Kcmil cables and will likely be responsible for the additional cost of the 600 Kcmil cables. As another example, if the contract contains a "no damage for delay" clause (i.e., the contractor's only remedy for delay is additional time, regardless of who causes the delay), the contractor submits a change order proposal which includes damages explicitly related to delay, and the owner pays those damages, it may well have waived (and eliminated) the no damage for delay clause.
- *What the owner writes* What the owner writes can modify contractual requirements. For example, if the design agreement contains a provision providing that the designer must

design to budget, if the owner writes to the designer instructing it to modify the design, and acknowledges that the modification might cause the cost of the project to exceed the budget (and doesn't require offsetting modifications), the owner may well have waived the design to budget requirement. As another example, if the design agreement requires the designer to complete all design work by a certain date, if the owner sends a letter establishing the date for completion of contract documents after the contractually established date for completion of all design work, the contractual deadline may very well have been waived.

- *What the owner says* What the owner says can modify contractual requirements. For example, if the contract requires a mock up to demonstrate paint colors, and the owner orally directs the contractor to paint the actual walls directly to save time, the contractual requirement to paint a mock up will be eliminated, and any extra costs resulting from the need to repaint the wall may be the responsibility of the owner. As another example, if the contract requires that change order work only be performed following the full execution of a written change order, and the owner orally instructs the contractor to perform certain work which the owner acknowledges as change order work without waiting for a written change order, the owner may well have waived the defense that subsequently performed extra work was performed on the basis of an oral conversation and not pursuant to a change order as required by the contract.

Each of these examples have been followed by the observation that the owner "may well have" relinquished some right under the design agreement or the construction contract. This wording is used because waiver situations, if ultimately resolved by a judge or arbitrator, are very fact specific. Therefore, whether a specific contractual requirement was waived depends on exactly what actually happened. With that qualification, the examples given are very likely to lead to a finding of waiver.

It is important to remember that the designer or contractor can waive contractual requirements and/or its position on issues as well. The most common instance of waiver by the designer or contractor is performing additional work (technically, additional services for the designer and change order work for the contractor) pursuant to an oral instruction by the owner under a contract that requires that all additional services be the subject of an amendment or a change order before the work is performed. If the designer or contractor proceeds on several occasions to perform additional work without a written amendment or change order, it may not be able at some later point in the project to refuse to perform additional work without a written amendment or change order.

The designer or contractor can waive its position on other types of issues as well. An example is provided by a project where the construction manager was contractually entitled to an incentive payment if it commenced driving piles by a certain date. The issue was whether the incentive was payable if the construction manager started driving indicator piles (used to test underground conditions) or production piles (to be used in the actual foundation) by the specified date. The construction manager claimed in meetings and in correspondence that the incentive applied to the driving of the indicator piles, which were the only ones it had started by the specified date. However, the construction manager's schedule showed the incentive applying when it started driving the production piles. The incentive was not paid. In this instance, the construction manager waived its stated position by its conduct; namely, the preparation of its project schedule.

The question then becomes how to avoid waivers. The owner should take five steps to achieve that objective. Exhibit 3.17 lists these five steps.

---

**Exhibit 3.17   Steps by owner to avoid waivers**

The owner should:

- draft an anti-waiver provision in each of its agreements and contracts;
- limit authority to give instructions under the contract to certain named persons;
- avoid acquiescing in procedures and/or actions inconsistent with contractual requirements;
- respond appropriately to correspondence and other communications with which the owner disagrees; and
- provide training to its project personnel on avoiding waivers.

---

The first step the owner can take is to seek to limit waivers through contractual provisions. Exhibit 3.18 is an example of such a provision.

---

**Exhibit 3.18   Anti-waiver contractual provision**

No act or omission of the Owner, or any of its authorized representatives, shall constitute a waiver of any provisions of the Contract Documents, even if there is an economic windfall to the Owner; provided that the Owner may waive any provision of the Contract Documents in a written document signed by an authorized representative of the Owner.

---

This provision can be used in the design agreement or consultant agreement simply by substituting "Design Agreement" or "Consultant Agreement" for "Contract Documents."

The second step the owner can take is to limit the number of persons who are authorized to give instructions to the designer or contractor. The fewer people authorized to give instructions, the less likelihood there will be of some member of the owner's team granting an unintentional waiver. Exhibit 3.19 illustrates two ways this can be achieved contractually.

---

**Exhibit 3.19   Contractual provisions limiting owner personnel authorized to give instructions**

The rights and responsibilities of the Owner under this Agreement [Contract] shall only be exercised by _____ [insert name(s) and title(s)]. No other person is authorized to give instructions to the Designer [Contractor] on behalf of the Owner.

The rights and responsibilities of the Owner under this Contract [Agreement] shall only be exercised by the person or persons specifically designated by the Owner. The Owner shall identify such person or persons in writing at the Preservices Meeting [Preconstruction Meeting].

---

The third step the owner can take to minimize waivers is to avoid approving or acquiescing in procedures and/or actions inconsistent with contractual requirements. This is really the key; if there is no approval of, or acquiescence in, actions that are inconsistent with contractual requirements, there can be no waivers. At a minimum, before agreeing, or failing to object, to a designer's or contractor's suggestion to proceed in a manner other than as required by the contract, the owner should carefully consider all the ramifications of the suggestion before approving or acquiescing.

The fourth step is responding to correspondence or other communications with which the owner disagrees. If the contractor sets forth in writing a position with which the owner disagrees, the communication must be answered as soon as possible and the disagreed with point(s) responded to directly. If this is not done, the owner may be held by a third party to have waived its position and conceded the contractor's point(s). The owner's position should be explained in as much factual and contractual detail as is necessary to communicate it persuasively. This provides the contractor full opportunity to reconsider its position, puts the contractor on notice as to the grounds for the owner's position, and provides the maximum chance that a third party will support the owner's position. Minutes of any meeting should be approached in the same manner. When the contractor takes positions with which the owner disagrees, they should be challenged, with a full explanation of the challenge offered. The minutes of such a meeting should be carefully reviewed to make sure the owner's position is correctly represented. The reason is the same; i.e., failure to challenge the contractor's position in a written record of a meeting may later be held to be a waiver of the owner's position on that issue.

The fifth step the owner can take to minimize waivers is to provide training for its project personnel on avoiding waivers. While the value of training applies to virtually all subjects covered by this book, there is a special value in providing guidance to owner personnel on the nature of waivers and how to avoid them. That is because the vulnerability to waivers is so pervasive because often owner personnel don't understand that they are waiving a requirement when they take (or fail to take) some action on the project, and because of the frequently recurring potential for waivers on most projects. Several final observations on waivers are in order.

- The concept of protecting against waivers is so vitally important because avoiding waivers really translates to protecting the owner's benefits under the design agreement, the consultant agreement, and the construction contract.
- All of the steps suggested to minimize waivers, even if combined, may not entirely stop waivers. That is because the key concept in the determination about whether the owner has waived a contractual provision is whether the designer or contractor reasonably relied on the owner's action, or failure to take action, and changed its economic position (i.e., incurred costs) on the basis of that reasonable reliance. If that occurred, it will be very difficult for the owner to avoid the consequences.
- On any given project, the owner may make a reasonable business decision to waive a provision on a one time or ongoing basis. Nothing in this section is intended to advise against such a decision, assuming it is taken deliberately and with an understanding of the potential consequences.

## Assembling and managing documentation

Because of the vital importance of documentation to the resolution of disputes, it is necessary that the owner have an effective system for assembling and managing project documentation.

This section will describe what constitutes the most important project documentation and offer some thoughts on how to approach managing the documentation.

There are four types of documentation that are particularly important to a construction project: the contract documents, documentation of what happens in the field, contractor submittals, and correspondence. Each of these is described below.

The contract documents involve the design agreement and the construction contract. The design agreement usually is a self-contained document. The scope of services are either included in the body of the agreement or in an exhibit. If there are other important aspects of the deal with the designer, such as a schedule for the completion of the various portions of the designer's scope of service, these provisions are typically included in the agreement as exhibits.

The construction contract typically consists of a number of parts. These include the owner–contractor agreement, the general conditions, any special and/or supplementary conditions, the specifications, and the plans. The documents frequently change as a result of adjustments made in the field and/or by change order. They should be conformed (i.e., made to reflect the latest changes) so as to have an accurate copy with which to administer the project. Also, a fully conformed contract (i.e., one containing all approved modifications) should be maintained.

For effective contract administration, it is crucial to have a record of what happens in the field on a daily basis. Daily reports provide an invaluable record to use in addressing contractor requests for change orders and/or claims. Exhibit 3.20 contains a list of the important information that should appear on a daily report.

---

**Exhibit 3.20    Important information for daily reports**

- Weather;
- which subcontractors are on the site;
- the number and trade categories of workers;
- the number and level of supervisors;
- what equipment is being used;
- what portions of the work are being performed;
- any problems or unresolved issues;
- pertinent comments by contractor/subcontractor personnel; and/or
- any unusual events.

---

There are several types of contractor submittals: (1) permits required to perform the work; (2) the project schedule; (3) shop drawings and samples; (4) requests for substitutions; (5) requisitions for payments and associated documentation; (6) warranties, manuals, and other documents required as a result of the contractor completing its work; and (7) various miscellaneous submittals required of the contractor by the contract. All of these should be assembled and retained.

It is important to keep all correspondence and to file it in a manner that every piece can be efficiently retrieved. Correspondence is the mechanism by which the inevitable contractual ambiguities and unforeseen problems are addressed and resolved. It is important to be able to assemble the written record as it relates to any specific issue that is in dispute.

Emails should be regarded as a form of correspondence. If an email relates to an issue that is or might be in dispute, or if it conveys information that will not be conveyed elsewhere, or if it is the only record of a transmittal or other action taken by a project participant, it should be printed out and retained in the same manner as a letter or memo. Regardless of the advent of the digital age, it remains critically important to be able to substantiate previous communication. When an email is used to communicate about a matter of substance, then a copy of that email becomes very important.

Comprehensive documentation requires an effective document management system. This system should include a standard procedure for project communications and a system for receiving, storing, and retrieving documents.

The owner has a particular interest in establishing a method for communicating that determines which people on the project will talk directly to which other people. This ensures that only those authorized to commit the owner are doing so and minimizes the chances that the owner will waive its rights or agree to extra work to which it did not intend to agree. A typical concern here is establishing a communications system by which the owner talks only to the contractor, and not directly to any subcontractors. This approach allows the owner to hold the contractor completely responsible for the satisfactory performance of all aspects of the work. For many projects, this procedure may not require more than a few paragraphs, but it should not be ignored, even on small projects.

The various types of documentation should be processed through a system that has two important objectives. The first is to ensure that no piece of paper or email of any potential importance is thrown away or misplaced. The second is to make any given piece of documentation easy to retrieve, which means a filing system that any member of the project staff would know how to use to find a piece of documentation.

The document management system should be set up to meet the needs of the owner and its personnel. However, all effective document management systems have several attributes in common. These are listed in Exhibit 3.21.

---

**Exhibit 3.21    Attributes of effective document management systems**

- *Comprehensiveness* All types of documents should be included in the system.
- *Internal logic* The major divisions and the subdivisions should be created on the basis of easily understood principles.
- *Ease of use* The system should be no more complicated than necessary to accomplish the objectives for each individual project. The instructions and/or key should be readily available.
- *Technologically current* The system should take advantage of currently available technology and should be upgraded as the technology advances.
- *Accountability* Someone specific should be responsible for the successful operation of the system. This includes ensuring that the requirements of the system are applied strictly and throughout the life of the project.

---

## Developing procedures

There are three reasons to develop written procedures. First, it requires the owner to carefully think through the myriad of issues involved in administering the design and construction of its projects. Second, procedures facilitate the management of projects by communicating roles, responsibilities, and expectations to project participants. Third, written procedures promote fairness and, if necessary, strengthen the owner's legal position by facilitating consistency (i.e., the treatment of similar situations similarly).

Well-written procedures should have at least three sections. The first section identifies the subject and purpose of the procedure. The second section establishes the objectives of the activity covered by the procedure. The third section describes how the activity will be performed, with particular attention to which positions are involved, what their respective roles and responsibilities are for this activity, what documents will be generated or reviewed, and how the documents will be used.

The procedures should identify responsible positions by title, not people by name. That is because the procedures are intended to describe how the owner as an organization is going to administer construction contracts. As an example, the project manager should perform the same functions on a specific project, regardless of whether the person in the position is Susan Smith or Frank Jones.

Owners may have different objectives for their procedures. Some owners may look to a procedure as a mechanism for providing greater detail on how a contractual provision (or set of provisions) will be implemented. Under these circumstances, the procedure would be distributed to the designer and the contractor. Among this group, some owners may wish to go into considerable detail, and make following the procedure a contractual requirement. Other owners may view their procedures as how their organizations will manage projects. These procedures would not be distributed to other project participants.

Exhibit 3.22, which illustrates one format for a procedure, is an example of a procedure intended for distribution to all project participants. It deals with administering the punch list.

---

**Exhibit 3.22    Sample procedure dealing with punch list administration**

**1.0  Purpose**

The purpose of this Procedure is to establish the steps for the preparation and resolution of punch lists (a listing of all known deficiencies or variations from the requirements of the Contract Documents) for the acceptance of the Work by the Owner.

**2.0  Objectives**

The Punch List Policy has the following objectives:

- *Contractual Basis* The Owner's consideration of the punch list shall be based on the requirements of the Contract Documents and is intended to ensure that the Owner receives the entire project described in the Contract Documents.
- *Fairness* The Owner shall seek to treat the Contractor fairly (e.g., treat similar situations similarly, acknowledge positions with merit promptly, etc.)
- *Efficiency* The punch list process shall be conducted as efficiently as possible so as to facilitate timely closeout of the construction contract.

**3.0  Procedures**

**3.1  Personnel**

3.1.1  The Owner's responsibilities under this procedure shall be carried out by the Owner's project manager.

3.1.2 The Contractor's responsibilities under this procedure shall be carried out by the Contractor's project manager, unless otherwise approved by the Owner.

3.1.3 The Designer's responsibilities under this procedure shall be carried out by the Designer's project level manager of construction administration, unless otherwise approved by the Owner.

## 3.2 Documents

The documents used in punch list procedures are as follows:

3.2.1 Written Request/Notification

The Contractor, the Designer, and the Owner shall document in writing all requests for inspection, modifications, description of findings, and any other communications related to the preparation modification, and completion of the punch list in writing.

3.2.2 Punch List

A written punch list shall be prepared by the Contractor and issued to the Owner. The Contractor shall be responsible for developing a punch list format that is acceptable to the Owner. Revised editions of the punch list shall be prepared as necessary. The final edition shall show all items satisfactorily resolved as described in Paragraph 3.3.4.

## 3.3 Process

The punch list process is described in this section.

3.3.1 Initiating the Process

When the Contractor believes it has substantially completed its work, it shall notify the Owner in writing and attach a copy of its proposed punch list. The Contractor will forward a copy of its request and punch list to the Designer.

3.3.2 Approving the Punch List

The Designer shall review the punch list as submitted by the Contractor and make any appropriate recommendations to the Owner to add or delete items. The Owner shall approve the punch list when it believes it to be complete.

The above steps shall be repeated until the Contractor has completed all of the punch list items. The Contractor shall then notify the Owner and the Designer, and request that they perform a final inspection.

3.3.3 Completing the Punch List

The Contractor shall complete and/or correct all items identified on the approved punch list and send a written request for re-inspection to the Owner with a copy to the Designer.

After being notified by the Contractor that the punch list work is complete, the Designer shall schedule an inspection. The Owner may participate or schedule its own inspection. Following such inspection, the Designer shall advise the Owner which items it believes have been completed as required by the contract documents, and which require further work. The Owner shall make the final determination as to whether items have been completed as required by the contract documents.

If there are items on the initial punch list that have not been completed as required, the contractor shall submit a revised punch list to the Owner, with a copy to the Designer, for the Owner's approval. The Designer shall make its recommendations to the Owner as to whether the revised punch list is complete. When the revised punch list is approved, the Contractor shall complete the items on the revised punch list. When it believes the revised list has been completed, the Contractor shall notify the Designer and request an inspection by the Designer and the Owner. If the Owner determines that there are still items that have not been completed as required by the contract documents, the steps described in this paragraph shall be repeated until all items on the punch list have been satisfactorily resolved as described in Paragraph 3.3.4.

3.3.4 Satisfactory Resolution of Punch List Items
Each item shall remain on the punch list until it has been satisfactorily resolved. An item on the punch list shall be satisfactorily resolved when the Owner determines that:

a  the work has been completed as required by the contract documents;
b  the Contractor has agreed to an acceptable credit for not completing the work as required; or
c  failure to complete the work does not reduce the value of the project and is otherwise acceptable.

The owner may not want to designate specific positions for the contractor and designer. In that case the procedure in Exhibit 3.22 would refer only to the contractor and the designer, without specifically referencing the contractor's project manager and the designer's project level manager of construction administration. A compromise would specify that the contractor and the designer will submit to the owner the name and position of the person responsible for coordinating the punch list process for their respective organizations.

If the procedure set out in Exhibit 3.22 was for internal use only, it might be more detailed in some respects and less in others. For example, if the owner's team included professionals in addition to the project manager, then the procedure might designate the project manager as ultimately responsible but state that the project engineer was responsible for tracking the work on the punch list to determine the status of the punch list; the mechanical, electrical, and plumbing (MEP) coordinator was responsible for inspecting all work related to the MEP systems; and a designated member of the user department was responsible for inspecting the finishes. The procedure might require a specified list of people to get copies of the punch list, including personnel in the user department. It might include a required format for the punch list, and a required form to transmit the punch list. On the less detailed side, it might describe the contractor's and designer's roles in only the most general of terms.

If the owner intends to supplement the contract documents with a detailed procedure for a certain project activity, for example the submittal of requisitions for payment, it should refer to the procedure in the contract documents. Exhibit 3.23 illustrates two possible approaches.

---

**Exhibit 3.23   Sample contract provisions to incorporate a procedure**

A provision covering payments to the contractor could read as follows.

> The Contractor shall prepare its requisition for payment in accordance with the Procedure for Payment of Requisitions attached to this Contract as Exhibit A and hereby incorporated into the Contract.

A provision in the design agreement and the construction contract could state the following.

> The Owner has a set of procedures for the administration of design and construction projects. Such procedures, entitled the Northern Hotel Company's Procedures for the Design and Construction of Hotels, are hereby incorporated into this [Design Agreement; Construction Contract]. By executing this [Agreement; Contract], the [Designer; Contractor] acknowledges that it has received such procedures, reviewed them, and understands and agrees to their requirements.

---

Project procedures should cover key contract administration activities. At a minimum, they should cover document management, review of requisitions for payment, change order and claims management, and contract closeout.

The previous section, Assembling and managing documentation, addressed document management. A procedure for this activity should begin by establishing the objectives of the document management system. These should include prompt distribution of incoming correspondence and documents, development of a comprehensive filing system, and accurate filing. The procedure should address which position receives incoming correspondence and other documents, which positions receive copies of the various types of project correspondence and other documents, and how documents are to be filed and retrieved. It should specifically discuss how electronic files are to be created, organized, and maintained.

The design agreement and the construction contract will address the responsibilities of the owner, the designer, and the contractor for submitting, reviewing, and approving requisitions for payment. The owner's procedure in this area will be aimed at providing a detailed process. The procedure should begin by stating the objectives of the procedure. These should include paying the contractor promptly amounts to which it is entitled, ensuring that requisitions are properly documented, confirming mathematical accuracy, and ensuring that completion percentages claimed by the contractor are reasonable. The procedure should identify the specific positions that are responsible for reviewing the requisition and what each review is intended to accomplish. For example, the controller might be responsible for determining mathematical accuracy, while the project engineer reviews the contractor's claimed completion percentages. It is important to determine that the percentage completion for the various activities claimed as payable by the contractor are realistic. This is because the percentage completion determines the amount payable and because lack of attention to this issue can easily lead to overpayment of the contractor at the early stages of the project with predictable declining interest on the contractor's part in fully completing the project. This procedure should also establish which position(s) will actually approve requisitions.

The procedure covering change order and claims management should begin by establishing the objectives for reviewing change orders and claims. These should include cost control for the owner, fairness to the contractor, and an efficient process. The procedure should describe the standards by which change orders and claims will be reviewed. For example, standards might include complete documentation as required by the contract (it is useful to specify the contractual requirements in the procedure) and a clear demonstration of entitlement. The design agreement and the construction contract establish the responsibilities of the owner, the designer, and the contractor for submitting, reviewing, and approving change orders and claims. This procedure should identify the specific positions that are responsible for reviewing change orders and claims, and the objective of each review. For example, review of change orders and claims should start with a review of entitlement; if there is entitlement, then, and only then, will it be appropriate to evaluate the contractor's proposed value for increased costs and/or time. The procedure should address which position(s) will analyze entitlement and which position(s) will review the contractor's estimate and schedule analysis. This procedure should also establish which position(s) will sign change orders. The procedure should discuss similar matters for analyzing claims, unless the same personnel and process is used for considering claims as is used for change orders. On projects of sufficient size, there should be separate procedures for change orders and claims because of the more routine nature of processing change orders as compared to analyzing claims.

The procedure for contract closeout should begin by listing the objectives of the closeout procedure. These should include ensuring that all contractual requirements have been satisfied, obtaining full documentation of the as-built project, and ensuring that all warranties have been issued as required and are received by the owner. The process should start with itemizing what contract deliverables are due prior to final payment. These may include such things as as-built drawings, a fully conformed copy of the specifications, warranties provided by the manufacturers of installed equipment, operations and maintenance manuals for large pieces of installed equipment, and insurance certificates evidencing that insurance policies required to be provided by the contractor for some period following completion of the work are in place. Of particular importance is evidence that all of the contractor's subcontractors and suppliers have been paid. This procedure should establish which specific position(s) is responsible for administering the contract closeout process. If there is more than one position involved, the responsibilities of each should be addressed by the procedure. The procedure should also identify the position responsible for determining that the process has been satisfactorily completed. In most cases, this should be the same position that is responsible for approving the final payment to the contractor.

## Using meetings effectively

There are three reasons why it is important to effectively manage project meetings. These reasons include achieving important owner objectives, preserving project morale, and not wasting people's time.

Project meetings are an important mechanism for the owner to achieve certain objectives. Meetings can assist the owner to establish policy. The people that need to be involved in the deliberations can be brought together. A meeting allows the participants to exchange views, and to benefit from hearing others' positions. The give and take of opinions by project participants gives the owner an opportunity to make decisions with whatever input it deems appropriate.

To obtain this input, the owner must give thought to who should attend the meeting. Attendees should include those who have responsibility for one or more of the issues being discussed, as well as those with relevant knowledge. Meetings whose major purpose is to consider issues should favor more rather than fewer participants. Meetings whose major purpose is to make decisions, on the other hand, should include fewer rather than more participants. That is because meetings to consider issues should allow for the broadest possible input, whereas meetings to make decisions should only involve those with relevant responsibilities.

Who is invited to attend meetings is also a morale issue. Various members of the owner's team may feel slighted that they have less responsibility, and/or that they have less standing than they believed if they are excluded from meetings that are ostensibly within their position's purview. Therefore, when persons with apparent responsibility in areas covered by the meeting are not invited, it may be advisable to explain to them why they were not invited. Notwithstanding comments to the contrary, most people prefer to be included in meetings they consider relevant to their positions.

The third reason that running meetings effectively is important is that on most construction projects, a great deal of time is spent on meetings. This means that poorly run meetings can waste enormous amounts of time. This, in turn, can lead to increased costs.

There are three rules for running effective (i.e., things get accomplished) and efficient (i.e., they get accomplished as quickly as possible) meetings. Exhibit 3.24 lists these rules.

---

**Exhibit 3.24    Rules for effective and efficient meetings**

Rule 1:  Determine the objective(s) of the meeting in advance.
Rule 2:  Develop a written agenda that leads to the accomplishment of the objective(s).
Rule 3:  Have a chair or facilitator of the meeting that keeps discussion focused on the meeting subject(s) and drives the discussion to required conclusions.

---

A brief explanation of each rule follows, starting with Rule 1. It is very important to establish in advance what a meeting is intended to accomplish. The usual objectives of design and construction meetings are sharing information (e.g., a meeting to look at preliminary schematic designs), monitoring progress (e.g., weekly construction progress meetings, a specific meeting called to determine why the commissioning process is behind schedule), and making decisions or developing recommendations for a decision maker (e.g., a meeting to determine if the test results on concrete test cylinders meet the contract requirements).

Developing a written agenda helps to keep the meeting discussion focused on the intended subjects. The agenda is the mechanism that tells all the meeting participants what the meeting is about and what topics will be discussed. The agenda establishes the relevant subjects for discussion. For that reason, the subjects listed on the agenda should be selected to further the meeting's objectives. For example, if the purpose of the meeting is to determine whether to withhold a portion of the designer's invoice, the agenda subjects should be limited to the issues that would be the factual and legal bases for the withholding.

The third rule, use of a leader to keep the meeting focused, is necessary for running an efficient meeting. To achieve the benefit, the meeting leader must be an active enforcer of the agenda. An effective leader will also keep the discussion focused on the issues and not let it become about the participants. Finally, he or she will make sure the discussion leads to the planned for conclusion(s).

## Identifying and winning designer and contractor games

Designers and contractors typically use certain strategies that can be considered "games." These games are played to achieve certain results. This section describes two games designers frequently play and four of the most popular contractor games, the way these games are played, the reason they're played, and how the owner should play the game.

The designer's favorite game is "We know best what design you should have." Frequently, the designer believes that it is uniquely qualified to determine the design of the project, and that it is more qualified than the owner to decide what the design should be. Moreover, the designer believes it is truly working in the owner's best interests. The way the designer plays this game is by incorporating its preferred approach in the plans and specifications, and arguing vehemently to the owner that, notwithstanding the owner's preferences and/or budget, this is the best approach. Some designers will raise the ante in this game by discussing their views with community groups and/or regulatory agencies without advance notice to the owner. In rare cases, designers will introduce modifications to the design without telling the owner.

The owner should play this game by doing at least two things. The first is by clearly establishing in the design agreement the owner's approval role in the preparation and distribution of the documents. The design agreement should preclude showing the documents to anyone other than the owner without the owner's written permission. The second thing the owner should do is require regular meetings to review the progress of the plans and specifications. These meeting should include substantive discussions about how the plans and specifications are progressing, how any plans or specs have changed from the last meeting, and the impact those changes have on the project.

The second game designers are prone to playing is the "Additional services" game. In this game, the designer views the basic scope of services as narrowly as possible and claims that virtually anything not clearly stated in the basic scope is an additional service requiring additional compensation. The intent here is to maximize revenue for the designer.

The owner should play this game by doing two things. First, the owner should carefully draft the owner–designer agreement to anticipate this problem. Chapter Four will deal with this drafting approach in detail. The second thing the owner should do is interpret the design agreement to further the owner's intent. That is, those things that were intended to be basic scope should be considered as such, and the designer should be actively challenged to the extent it seeks, by contract interpretation, to expand the opportunities for additional compensation.

The first contractor game is "Bid low and make or increase profit through change orders." This game is the contractor's equivalent to the designer's "Additional services" game. The contractor bids low to get the work, and then treats any response to a request for information, submittal, or instruction of any kind by the owner as a change order requiring extra compensation (and, on occasion, extra time). The objectives of this game are simultaneously to obtain work and maximize profits on that work.

The owner should play this game by doing four things. First, the owner–contractor contract should be carefully drafted to create clear and rigorous standards for change order proposals not the result of owner-directed changes. Chapter Seven will address this issue in detail. The second thing the owner should do is to perform field inspections frequently, daily if possible. The third thing the owner should do is carefully analyze all contractor proposals for merit.

The fourth thing is, on proposals of any size that have merit, the owner should perform an independent evaluation of the value of the proposal. Chapter Nine discusses the evaluation of change orders in detail.

A key point in terms of contesting this game successfully is to establish very early in the project that the owner will make obtaining additional compensation through change orders a difficult and demanding process. That will do two things. First, it will discourage excessive number of proposals, and, second, it will leave the contractor satisfied with lower settlements on negotiated resolution of proposals.

The second contractor game is "We need a decision now." There are two objectives of this game. The first is to force the owner to make a speedy decision, either to accommodate what the contractor honestly believes is a time crunch, or to pressure the owner to make a decision without the contractor having to take the time and trouble to submit the information the owner (and, for some issues, its designer) really needs to make the decision. The second objective is to shift the responsibility onto the owner for any delay that may result from any time the owner takes to make the decision.

The owner should do three things to contest this game. The first is to draft the construction contract to clarify the owner's right to obtain any documentation it reasonably needs to make decisions. The second thing the owner should do is to insist on its rights to receive the information necessary to make a reasoned and informed decision. The third thing the owner should do is to document requests for the information necessary to make the decision. This record of requests, particularly if it includes multiple requests, will provide an effective response to any claim for delay by the contractor based on the length of time taken by the owner to make a decision. The request should clearly explain what information is sought and should reference any previous requests. It should also specifically preclude the owner's responsibility for any resulting delay. Exhibit 3.25 illustrates such a request.

---

**Exhibit 3.25    Request for information necessary for a decision**

Widget Manufacturing Company
125 South Street
Anytown, State 01112

March 26, 2016

XYZ Construction Corp.
10 Main Street
Anytown, State 01111

Attn: Ms. Susan Smith; Project Manager

**RE: Construction of Riverside Facility / Proposed Substitution for Conference Room Windows / Request for Additional Information**

Dear Susan:
This is in reference to XYZ's request for Widget's approval for a proposed substitution of Wilson Windows Model No. 2234-CTF (the "Wilson") for the specified Mercury Windows Model No. 18-CTF (the "Mercury").

   XYZ initially submitted its request for substitution of the Wilson windows for the originally specified Mercury windows on November 23, 2015. On December 3,

Widget responded by sending a letter to XYZ requesting additional information on the characteristics of the proposed windows. That request has been discussed at every progress meeting since then. Additionally, on January 4 and February 1, 2016, Widget renewed its request for additional information.

As indicated in Widget's previous correspondence, both the originally specified Mercury and the proposed Wilson are floor to ceiling interior windows for the conference room in the executive office area on the third floor. The specifications required that these windows have certain characteristics related to sound transmission and resistance to scratching. To date, XYZ has not submitted any information concerning these characteristics of the Wilson windows.

Any delay associated with or resulting from this requested substitution by XYZ shall be the sole responsibility of XYZ. Furthermore, if the requested information is not received by Widget by May 15, it will consider that XYZ has withdrawn its request for substitution.

Sincerely,

Frank Jones
Project Manager

The third game contractors play is called "Not performing work to let schedule force the desired decision." The purpose of this game, from the contractor's perspective, is to force the owner to make a decision in the way that the contractor wants it made, but which the owner is resisting making. For example, the owner may want to upgrade the cable that brings power to a conference room in order to be able to have sufficient power to enable conference room participants to use the latest electronic devices. The owner may not dispute that this is an extra. However, to implement this change, the components of the main building power switch have to be changed, and the contractor is concerned that these changes will adversely impact the schedule and, perhaps, preclude obtaining a certificate of occupancy by the contractually required date. The contractor may simply not perform the upgrade, let the drywall contractor close up the walls in the conference room, and then confront the owner with the disproportionate extra cost now involved in upgrading the cable.

This game is not only played with respect to owner-directed changes. It is also frequently played with respect to original scope work. For example, if there is a wood ceiling required in the same hypothetical conference room, the specifications may establish certain testing requirements to ensure that the panels are properly fire retardant. The contractor may wish to substitute different panels than those originally specified. The contractor's proposed panels may be of lesser quality or have less aesthetic appeal, but they may have an established fire rating. The owner may regard them of inferior quality and prefer its originally specified panels. The contractor may delay submitting the originally specified panels for testing, intending to tell the owner that failure to approve its proposed panels will significantly delay the project.

The owner should contest this game by doing two things. First, the owner should have an accurate picture of how the decision it has to make is going to impact cost and schedule. This information will be available if the owner has good site inspection, cost reporting, estimating, and schedule analysis capacity. Second, the owner should not be reluctant to provide explicit instructions to the contractor at an early stage concerning the approach the owner wants taken, thereby offsetting the contractor's ability to use the passage of time as a point of leverage. Exhibit 3.26 illustrates such an instruction.

**Exhibit 3.26   Instruction on performing work to avoid delay**

Major City Bank
1 Bank Plaza
Major City, State 04128

March 16, 2016

Atlantic Construction Management Co., Inc.
23 Day Boulevard
Major City, State 04128

Attn: Mr. David Wilson; Project Manager

**RE: Major City Bank Office Building / Completing Executive Conference Room**

Dear Dave:

This in reference to the testing and installation of the wood panel ceiling in the executive conference room in the new office building ("the Building") owned by Major City Bank ("the Bank") and to the request by Atlantic Construction Management Co., Inc. ("Atlantic") to substitute a different wood panel for the ceiling.

Paragraph 2.04(B) of Specification section 10300 describes the wood panel that is to be installed in the ceiling of the executive conference room. Paragraph 2.05(A) requires that samples of such panels be submitted to a testing agency selected by Atlantic and approved by the Bank to confirm that it has the contractually required fire rating.

Failure by Atlantic to submit the necessary samples for testing is raising a serious possibility of delay to the schedule for completing the Building. To complete the testing, receive the test results, complete the evaluation, and then order the necessary number of panels and receive them on site so as not to delay the job, requires that the testing be completed within the next week. The need to submit sample panels for testing has been discussed at the last three progress meetings.

The Bank is aware that Atlantic has suggested using a different panel which it has recently used on another project and which obtained the necessary fire rating. However, for reasons involving the policy for Building finishes, and particularly for the executive conference room, the Bank does not wish to consider the proposed substitute unless the originally specified panel fails its testing requirements. Incidentally, the Bank is aware of at least two other projects in Major City where the originally specified panels have been used and have not been a problem for the building or fire department inspectors.

Atlantic is instructed to immediately submit the contractually specified wood panel samples for the contractually required testing. Any delay to the completion of the Building caused by Atlantic's timing in submitting the panels for testing will be the responsibility of Atlantic, and any damages to the Bank resulting from such delay shall likewise be the responsibility of Atlantic as set forth in Paragraph 7.2 of the General Conditions.

If you have any questions, please call me.

Sincerely,

James MacDonald
Vice President—Capital Facilities

The fourth game contractors play is "We work for you." Contractors have become fond of telling owners, "We work for you," or "We have your best interests in mind." There are two objectives of this game. First, the contractor wants to create a climate in which the owner is less likely to question the contractor's positions throughout the duration of the project. Second, in connection with a specific issue, the contractor will use this as a way of trying to obtain the owner's agreement on an issue without much of any analysis.

The owner should not contest this game. Rather, it should refuse to play. The owner should simply treat these statements as rhetoric and ignore them. In most cases, the statements are meaningless. Those contractors whom the owner truly believes are working in its best interests will not need to make statements of this kind because they will demonstrate it with their actions.

## Not pointing an unloaded gun and other miscellaneous techniques

This section considers four miscellaneous techniques of contract administration. These include not pointing an unloaded gun, documenting all observations and conversations, not leaving allegations and accusations unanswered, and not minimizing the importance of emails.

### *Don't point an unloaded gun*

Pointing an unloaded gun refers to threatening to do something one knows in advance one is not prepared to do. The owner threatening to withhold payment to the contractor for failure to remove construction debris as required by the contract when the owner is not prepared to withhold payment for that reason is an example of pointing an unloaded gun.

There are three reasons for not making threats one has no intention of carrying out. The first and most important is that it erodes credibility. While doing it once may result in a successful bluff, a pattern of unfulfilled threats, which is the more usual pattern, will cause the contractor to ignore most owner attempts to use leverage unless and until the owner actually follows through on some threatened action.

The second reason not to point an unloaded gun is that it may raise the tension level of the relationship without corresponding benefit. To threaten the designer or the contractor with an action adverse to its interests will run the risk of irritating the designer or contractor. To do that when there is no intention to follow through is to create tension for insufficient reason.

The third reason not to point an unloaded gun is that the designer or contractor may take the threat as real and dig in its heels. For example, if the owner threatens to withhold partial payment till the designer completes the application for a permit required for the project, the designer may respond by not completing the application on the theory that it should get paid for at least the effort it has put into the application to that point. If there is no intention to withhold, the threat will have created a response that worsens rather than improves the situation from the owner's perspective.

### *Document observations and conversations*

It is important to create a record of all observations and conversations of any potential importance. Conversations that deal with pleasantries or other inconsequential matters need not be documented. Those that deal with issues that are, or may become, in dispute are essential to document. The reason documentation is so important is because things that the designer or the contractor says or does can often be used to contest a claim for additional compensation or additional time.

The Understanding and avoiding waivers section in this chapter discussed how what the owner does, says, or writes can alter the terms of the contract as the contract is being performed. What the designer or contractor does, says, or writes can undercut a proposal or claim for additional compensation or time. For example, the contractor is alleging delay because the plumbing subcontractor was late in completing its work because the owner elected to change bathroom fixtures. The contractor's superintendent, in an informal conversation, acknowledges the plumbing subcontractor was late because it did not have enough plumbers performing the work. This statement seriously undercuts the contractor's claim for delay and should be carefully recorded for use at the appropriate time.

### *Don't leave allegations and accusations unanswered*

The importance of considering third parties when crafting correspondence because they can impose resolution of disputes was addressed in the Preparing effective communications section. Because of the potential involvement of third parties it is important that the owner respond to allegations or accusations with which it disagrees. Failure to respond creates the risk that a third party will interpret the lack of response as acquiescence.

An example will illustrate. The design agreement says the designer "will assist the owner to obtain all permits necessary for the commencement of construction." The owner and the designer attend a permit hearing and present a drawing depicting the project. The permitting authority requires that a second drawing be prepared to support the permit application. The owner requests the designer to prepare the second drawing. The designer agrees, stating in an email, "Our obligation to assist is not open ended. We have already prepared one drawing for the [name of the public authority] so this second drawing will be an additional service."

The owner will need to respond with its own email. The response should be:

> Paragraph [number] of the design agreement requires [name of designer] to assist with all permit applications. That provision does not limit the obligation to a single drawing. The preparation of an additional drawing required to obtain the permit is not an additional service.

Failure to write such a response, if the issue goes to a third party, may be held to be acquiescence by the owner in the designer's position that it is an additional service.

Two temptations should be resisted in connection with designer or contractor assertions with which the owner disagrees. The first is to respond with anger and/or incredulity and then conclude that the assertion is so absurd it does not deserve a response. There is no guarantee a third party would consider the assertion absurd, and the failure to respond creates the same risk that the third party will view the lack of response as acquiescence.

The second temptation is to not follow up a conversation about the issue. Often the owner will seek to resolve an assertion with a phone call or a meeting. This usually makes sense as a next step. However, unless the designer or contractor withdraws its assertion in writing, the assertion remains unanswered on the record, and, if the issue is not fully resolved and goes to a third party, the owner risks being found to have acquiesced in the original assertion.

In the example above, the owner might call the designer to discuss the issue of the second drawing. If the issue is resolved in the phone call or meeting, the owner should respond with an email that says, "Confirming our telephone discussion of this afternoon, [name of designer] agrees that the preparation of a second drawing will not be considered an additional service." If the issue is not resolved, then the owner should respond as suggested above, citing the appropriate provision of the agreement and stating the owner's view that the second drawing is not an additional service.

### Don't minimize the importance of emails

In today's electronic world, much project communication takes place by email. Emails are often written more informally than letters. However, that doesn't change the potential importance of any individual email, or emails in general. The emails sent and received by owners should be written and retained using all the recommendations in this book.

## Minimizing disputes

Resolving issues before they become disputes is in the interests of the owner, provided the cost and schedule aspects of resolution are reasonable. One test of reasonableness is whether the cost of resolving the issue is less than the cost of resolving it after it becomes a dispute. When estimating the costs of resolving a dispute, it is necessary to include transaction costs, including such costs as fees for consultants and attorneys.

There are at least three techniques for preventing issues on which there is a need for resolution from becoming disputes. These are resolving issues promptly, agreeing to meritorious positions, and firmly opposing positions without merit.

### Resolve issues promptly

Decisions, interpretations, and approvals (or disapprovals) should be issued as quickly as possible. This will avoid or reduce impacts on costs and/or schedule. It will often be perceived by the designer and the contractor as evidence of the owner's good faith. However, the owner should resist pressure to resolve issues prematurely. "As quickly as possible" means as quickly as possible after the owner has received all information and/or documentation necessary to make the decision. For example, if the designer presses for approval of the schematic designs without submission of a contractually required estimate the owner should insist on receiving the estimate prior to acting on the schematic design. Similarly, if the contractor urges the owner to approve a substitution without the contractually required statement of costs associated with the substitution, the owner should resist making the decision till it receives the cost information. On the other hand, not receiving unimportant information should not be used as a reason to delay a decision for which adequate documentation has been supplied.

### Agree to meritorious positions

The owner should agree readily to positions taken by the designer or the contractor that the owner considers meritorious. This too will reduce potential adverse cost and schedule impacts and will be seen by the designer and the contractor as evidence of good faith.

### *Oppose positions without merit*

Contractor positions viewed by the owner as unjustified should be clearly and firmly opposed. When done promptly, and in combination with the first two techniques, this approach can be very effective in resolving issues before they become disputes. This is because the contractor will come to understand that the owner's opposition is real rather than reflexive. At that point, the contractor will spend significantly less time and resources testing the owner's resolve.

# 4 Key issues in drafting the designer agreement

## Introduction

The owner–designer agreement is a critical document for the owner because it governs the relationship between the owner and the designer. That is why it is so important that the design agreement be drafted to fully protect the owner's interests. This chapter assumes the project participants are the owner, the designer, and the contractor; and that the owner is relying on its designer for the administration of the construction contract. If the owner has significant in-house project management capacity, or if the owner procures a third party project or construction manager, certain of the provisions may become unnecessary.

The owner should give careful consideration to the degree of its own competence to manage the project, particularly during the construction phase. If the owner does not have experienced construction professionals on its staff, it will need to make sure that it requires the designer to provide the necessary assistance. Therefore, the extent of the designer's responsibility for such things as construction inspection, schedule administration, review of change orders, and analysis of claims should reflect the extent to which the owner looks to the designer to provide project management services to supplement its own management capacity.

On some projects the question arises as to whether engineering disciplines and/or other design-related disciplines should have direct agreements with the owner as opposed to being subconsultants to the architect. Some owners believe they have greater control of the design process if all design disciplines are contracted directly to the owner. Some engineering firms prefer this arrangement because they believe architectural firms are not good at administering subconsultant agreements. In their view this includes slow payment and slow response to contract administration issues.

Notwithstanding these possible benefits, it is recommended that there be a prime designer on each project and that all other firms providing design services be subconsultants to the prime design firm. This is because the key issue from the owner's standpoint is coordination of design disciplines. Coordination of design disciplines refers to the process of ensuring that all aspects of a project's design relate to each other appropriately. For example, a building's heating, ventilation, and air conditioning (HVAC) system is powered by electricity. The project's plans and specifications have to address how power is brought to the HVAC system. This requires coordination between the mechanical and electrical design disciplines. Similarly, both of these systems (i.e., mechanical piping and electrical conduits) may have to pass through structural members. This requires coordination between mechanical, electrical, and structural design disciplines.

If all design disciplines work as subconsultants to the prime designer, then the prime designer has the responsibility for ensuring coordination between the design disciplines.

If each design discipline is contracted directly with the owner, the owner will be responsible for coordination of the design disciplines. The prime designer is significantly better equipped technically to carry out coordination of the design disciplines than is the owner. Furthermore, if there is a failure of coordination, the owner will want to be able to look to the prime designer's professional liability insurance.

A complete design agreement is included in the Appendix section at the end of the book as Appendix A. This document has been drafted to serve owner clients. It is closely related to the construction contract discussed in Chapter Seven and included as Appendix C. The paragraph numbers that appear in certain of the sample provisions are paragraph numbers from the sample design agreement.

## Major issues for design agreements

There are a number of key issues in drafting the designer agreement that can significantly impact the successful completion of the project. These issues can be divided into two groups: issues related to the delivery of the designer's services and issues related to the business deal between the owner and the designer. The issues in both groups are identified below and then discussed individually.

The following are important issues related to the delivery of design services.

- *Designing to the owner's program* The agreement should require the designer to prepare plans and specifications that faithfully implement the owner's program.
- *Designing to the owner's budget* The agreement should require the designer to prepare plans and specifications that produce a winning construction bid that is within the owner's budget, and, if no bids meet the budget, to assume the financial responsibility of the necessary redesign work.
- *Providing complete design services* The agreement should require the designer to provide a complete set of design documents that describe all of the work involved in the project and provide all other services necessary to perform the scope of services.
- *Responsible for subconsultants* The agreement should provide that the designer is responsible for the acts and omissions of its subconsultants.
- *Construction inspection* The agreement should require the designer to perform the level of construction inspection appropriate to its overall responsibility for assisting the owner to manage the project.
- *Schedule administration* The agreement should require the designer to provide schedule administration services.
- *Reviewing requisitions for payment* The agreement should provide for a full review of contractor requisitions for payment that goes beyond checking mathematical accuracy and involves reviewing the schedule of values with a focus on the claimed percent complete of the various activities, as well as reviewing other accompanying documentation.
- *Evaluating change orders and claims* The agreement should require the designer to play an active role in the evaluation of change orders and claims.

The following issues relate to the business deal between the owner and the designer.

- *Reimbursable expenses/additional services* The agreement should preclude reimbursable expenses except when required and approved in advance by the owner. Additional services should be limited, with the agreement making clear that key services such as construction inspection, schedule administration, and evaluation of change orders and claims are part of basic services. Reimbursable expenses are discussed in more detail below and additional services are discussed in detail in Chapter Five.

- *Ownership and use of documents* The agreement should provide for unfettered use of the documents by the owner, including continuing use as necessary for maintenance, rehabilitation, and/or expansion of the facility.
- *Payment provisions* The agreement should provide when the designer will be paid, what information the designer's invoices must contain, what documentation must be attached, and under what circumstances the owner may withhold payment.
- *Schedule for delivery of services* The agreement should establish when major design activities will be started and completed.
- *Deliverables required for management* To manage the designer, it is necessary that the designer produce certain deliverables against which its performance can be measured, in particular a monthly report. If these deliverables are not set out in the design agreement, their submittal will become an additional service.
- *Insurance* The agreement should establish the amount of errors and omissions and other types of insurance that the designer will be required to maintain. It should also specify the important requirements of each policy.
- *Standard of care and indemnification* The standard of care is the degree of care the designer owes the owner in performing its scope of services under the owner–designer agreement. Damages sustained by the owner as a result of acts or omissions that violate the standard of care are generally recoverable by the owner. For that reason establishing the relevant standard of care for the project is very important. The indemnification provision addresses the extent of the designer's exposure for the damages resulting from its errors and omissions. This issue and suggested agreement wording is discussed in detail in Chapter Five.
- *Termination* The agreement should describe the circumstances under which the owner can terminate the design agreement. This should include termination for convenience as well as termination for cause.

## Actual agreement provisions

The balance of this chapter discusses how to draft specific provisions that address the listed issues.

### Designing to the owner's program

The owner designs and builds a project to meet specific needs. These needs will address various aspects of the project as shown in Exhibit 4.1.

It is important that those needs be conveyed to the designer in sufficient detail that the designer can understand them and produce a set of plans and specifications that will result in the project the owner set out to obtain. It is not unusual for an owner and a designer to have somewhat different conceptions of what would make a successful project. For that reason, the requirement to design to the owner's program should be clearly stated in the design agreement. That requirement might be worded as shown in Exhibit 4.2.

---

**Exhibit 4.1    Issues addressed in owner's program**

- *Use* Will the facility be used as a laboratory, a manufacturing facility, a warehouse or an office building? If the owner is a public entity, will the facility be a school, a police station, a bridge, or a city or town hall?
- *Function* Within the general category of use, what is the function of this particular facility? For example, will an office building be the corporate headquarters, a regional office, a combination of office and backroom activities? Will a school be an elementary school or a high school?
- *Aesthetics* Does the organization have design standards that all buildings must meet? Are there marketing considerations that impact the design?
- *Community relations* Are there permitting requirements that impact the design? Are there aspects of design related to the organization's commitment to being a good neighbor?
- *Cost and schedule* Every project will have significant constraints related to cost and schedule.

---

**Exhibit 4.2    Designer's obligation to design to owner's program**

2.9  *Owner's Program* By submitting its Proposal, Designer represents that it has fully familiarized itself with the Owner's program, its objectives for this Project, and the Owner's Budget described in Paragraph 5.4. Failure to so familiarize itself with such program, objectives, and Owner's Budget will not reduce Designer's responsibility for meeting such program and objectives, and for preparing plans and specifications which allow the Project to be built for an amount no greater than that set forth in Owner's Budget.

---

### Designing to the owner's budget

Every project will have a budget. It is vitally important that the cost of the project not exceed the budget. However, some designers will increase the scope of the project because, in the designer's opinion, a successful project requires such things as upgraded exterior and/or interior finishes, mechanical systems, and/or more sophisticated low voltage (e.g., cable, wireless) systems. This inclination to "improve" the project for the owner actually impacts not only scope but potentially cost as well. That is why it is important to require the designer to design to the owner's budget.

To achieve that objective, the design agreement should do three things.

1    *Establish the designer's obligation* The agreement should obligate the designer to design the project within the owner's budget as illustrated in Exhibit 4.2.
2    *Require estimates during design* To make sure that the designer is focused on meeting the owner's budget, the agreement should require the designer to produce cost estimates at each stage of design (i.e., schematic design, design development, and construction documents).

3    *Responsibility for redesign* The agreement should establish that the designer bears the
     financial responsibility for any redesign work required to obtain a bid that complies with
     the owner's budget.

The following provisions in Exhibit 4.3 address the latter two points.

---

**Exhibit 4.3    Designer's obligation to design to owner's budget**

3.8    *Cost Estimates* Based on the Owner's Budget provided pursuant to Paragraph 5.4,
       Designer shall provide cost estimates to Owner as part of the submission of
       Schematic Design Documents, Design Development Documents, and Construction
       Documents. Each such estimate shall be broken down by trade. Each such cost
       estimate shall be approved by Owner as part of its approval of the Schematic,
       Design Development, and Construction Documents.

3.9    *Exceeding Cost Estimates* If Designer has reason to believe that the cost estimate
       to be submitted with the next set of Design Documents is at least five percent
       greater than the amount approved by Owner with the previous set of Design
       Documents, Designer shall inform Owner of such projected increase of cost
       immediately upon concluding such increase is likely. Owner and Designer shall
       consider appropriate alternatives, including, but not limited to, increasing the
       Owner's Budget, modifying the design, and/or changing materials and equip-
       ment. Any modifications to the Design Documents made because of such pro-
       jected increase in costs shall be performed at no additional cost to Owner unless
       such modifications result solely from changes to the Design Documents directed
       by Owner for reasons other than meeting Owner's Budget.

3.10   *Owner Approval for Increasing Budget* In no event shall the Owner's Budget be
       increased unless approved in writing by Owner. If Designer believes it is
       necessary to exceed the Owner's Budget, it shall submit to Owner a written
       request for approval that shall include an explanation of the reasons for
       the request, a description of Designer's efforts to avoid the request, and the new
       proposed amount for the Owner's Budget. Owner shall respond to such request
       in writing by either rejecting such request or by approving it at the requested
       amount or at some alternative amount determined by Owner.

3.13   *Lowest Construction Bid Exceeds Budget* If the value of the lowest construction
       bid exceeds the Owner's Budget, Owner at its sole discretion, shall elect to
       increase the Owner's Budget, rebid the Project, terminate the Project, or order
       the redesign of the Project to meet the Owner's Budget. If Owner instructs
       Designer to redesign the Project because the lowest bid exceeded the Owner's
       Budget, such redesign shall be accomplished at no additional charge to Owner.
       Owner shall participate in such redesign to the extent necessary to ensure that
       such redesign meets Owner's objectives for the Project.

---

### Providing complete design services

The designer should be obligated to provide a complete set of design services. All the
elements necessary to build the project should be clearly and comprehensively described and

illustrated. The plans and specifications should be in compliance with all applicable laws, regulations, and codes.

The provisions in Exhibit 4.4 illustrate how to create this requirement.

---

**Exhibit 4.4   Providing complete design services**

3.1 *Design Documents* Designer shall prepare all plans and specifications (collectively, "Design Documents") necessary for the construction of the Project. Such Design Documents shall be complete, in sufficient detail to allow the contractor selected by Owner ("Contractor") to construct a complete Project and to construct the project in full compliance with all applicable federal, state, and local laws, regulations, ordinances, and building codes.

3.2 *All Necessary Services* Designer's services shall include all architectural, engineering, construction administration, and other services necessary to perform the Scope of Services.

---

### *Responsible for subconsultants*

The prime design consultant (generally, an architect for buildings and an engineer for highways, water and sewer projects, and other horizontal projects) should be responsible for all the acts and omissions of its subconsultants. There are two reasons for this type of provision. The first is administrative. In most cases, the owner does not want to deal with the 2 or 3 (small buildings and horizontal projects) to 30 or more (large, complicated buildings) subconsultants on the project. The owner hires the prime consultant expecting it to manage its subconsultants.

Managing in this context has design service and business aspects. From the design services standpoint, the prime consultant needs to make sure that all the required disciplines perform their services comprehensively and on time. A key responsibility of the prime consultant is ensuring that the design documents are fully coordinated; that is, that the design documents describe and illustrate the needs of all trade categories in a manner that allows for the construction of the project. An example is the need for structural drawings to accommodate attachments and penetrations for mechanical and electrical piping and conduits.

From the business standpoint, management means administration of the subconsultants' agreements. This includes such activities as drafting an agreement that is consistent with the requirements of the prime agreement, paying the subconsultants appropriate amounts on a timely basis, and resolving subconsultant requests for additional compensation.

The second reason for this type of provision is to ensure that the prime design consultant is legally responsible for the services provided by its entire team. The owner has hired the prime design consultant to provide a complete set of design documents (and other deliverables in the design agreement). The owner should be in a position to hold the prime design consultant responsible for providing all the deliverables required by the design agreement and not have to pursue each discipline separately.

Exhibit 4.5 illustrates how to deal with both of these concerns.

---

**Exhibit 4.5   Prime design consultant responsible for subconsultants**

2.1 *Designer's Services* Designer's services shall consist of all services required by this Agreement, whether performed by Designer, its subconsultants, or any other person or entity performing services on behalf of Designer. Designer shall be responsible for all the acts and omissions of its subconsultants and any other person or entity performing services on behalf of Designer.

3.3 *Coordination* Designer shall take all necessary steps to ensure that the preparation, completion, and administration during construction of the Design Documents are fully coordinated. Coordination for the purposes of this Paragraph shall mean the timely and appropriate involvement of each of the various design disciplines throughout the design and construction of the Project.

---

*Construction inspection*

One key determinant of a successful project is effective construction inspection. Such inspection is vitally important to determine if the contractor is performing the work as required by the construction contract. The inspection, to be meaningful, must have three features.

1   *Frequent* The work must be inspected frequently; for a project of any size, this means daily.
2   *Detailed* The inspection must involve a serious review of all of aspects of the work.
3   *Focused on defective work* The inspector must have an affirmative obligation to identify and report defective work to the owner.

The typical standard design agreement requires the designer to make periodic visits to the site to observe the progress of the work and to determine if the work complies generally with the intent of the contract documents. If the designer should happen to observe defective work, it will report it to the owner. This level of construction inspection is not sufficient to protect an owner that is not performing frequent, detailed construction inspection itself. Exhibit 4.6 illustrates wording that requires an appropriate level of construction inspection.

---

**Exhibit 4.6   Construction inspection**

3.19 *Inspection of the Work* Designer shall regularly inspect the progress of the Work. Such inspections shall be daily, unless some other interval is approved in writing by Owner, and shall be of sufficient duration to provide Designer with detailed knowledge of the progress of the Work. Immediately following each such inspection, Designer shall complete a field report in a form approved by Owner. Copies of such field reports shall be attached to Designer's Monthly Report described in Paragraph 3.25, except that a copy of any field report which describes an actual or potential significant problem shall be provided to Owner no later than the end of the day on which the inspection was conducted.

---

*Schedule administration*

In virtually every case, the owner is concerned that its project be completed on time. For this reason, it is important to require the contractor to submit a schedule and to update it regularly. If this obligation is to have any real impact on the contractor, these schedule submissions must be reviewed and approved.

If the owner does not have in-house capacity to review schedules and has not hired a project management consultant, it becomes necessary for the designer to have this capacity. From the owner's standpoint, it doesn't matter if the person doing the schedule review is the designer's employee or a subconsultant.

It is important to understand the extent of the approval required. On the one hand, the contractor's schedule should represent a realistic plan for completing the work by the contractually required completion date. On the other hand, the owner does not want to approve the contractor's means and methods and/or sequencing of the construction because that may expose the owner to extra costs associated with that level of involvement in the project planning. Exhibit 4.7 shows wording which establishes the designer's obligation to review and approve the contractor's schedule submissions, while establishing appropriate limits on the approval function.

---

**Exhibit 4.7    Review of contractor's schedule**

3.18   *Review of Contractor's Schedule* Designer shall review and recommend approval or disapproval to Owner of the baseline schedule and schedule updates required by the Construction Contract. Such approval shall be limited to determining that such schedule submissions provide a realistic approach to completing the Project within the Contract Time, as defined in the Construction Contract, and that the updates also accurately reflect the progress of the Work.

---

*Reviewing requisitions for payment*

Effective review of requisitions for payment should include more than just checking for mathematical accuracy. It is important that the contractor not "get ahead" of the owner (i.e., when the value of payments to the contractor to date exceeds the value of the work performed to date). This is because when the contractor "gets ahead" of the owner in this sense, the contractor has various incentives not favorable to the owner's interests. The most common example of this is at project closeout. Contractors "getting ahead" of owners is one of the reasons owners have problems getting their contractors to fully perform the punch list work and to finally complete projects, including the paperwork aspects of closeout such as submittal of as-built drawings, warranties, and lien releases.

Contractors will frequently seek to front end load their jobs (i.e., try to get ahead) in order to ensure a steady cash flow, increase their profit (with interest on cash they don't have to use), and/or fund work on other jobs. To avoid the contractor getting ahead, a careful analysis must be done of the contractor's claimed percentage of the job completed because it is the percentage of work complete that drives the amount of the contractor's requisition.

The amount of the contractor's requisition is determined based on the schedule of values. The schedule of values requires the contractor to break the project down into a series of activities and to assign a value to each activity. Subsequent columns call for, among other information, the value of the work performed to date for each activity, the amount paid to date for each activity, and the percent complete of each activity. The amount due to the contractor for any given requisition is the difference between the total value of the work performed to date minus the total amount paid to date to the contractor. It is the percent complete that establishes the total value performed to date and, thus, ultimately, the amount due for the current requisition (the amount paid being an objective, documentable value).

An example may be helpful. Consider the construction of an office building. Assume the foundation item in the schedule of values is $800,000, and that the contractor has been paid $150,000 for the foundation work prior to the current requisition. Assume further that the contractor claims that the foundation work is 40 percent complete, and the designer concludes that the work is 30 percent complete. Exhibit 4.8 shows how the difference in percent complete leads to significantly different approvable amounts for the contractor's requisition.

---

**Exhibit 4.8    Calculation of amount due contractor**

| | | |
|---|---|---|
| Schedule of values total value for foundation work: | | $800,000 |
| Contractor's claimed percent complete (40%): | 0.40 | |
| Total amount due contractor to date: | | $320,000 |
| Total paid to date: | | $150,000 |
| Amount of contractor's invoice: | | $170,000 |
| | | |
| Schedule of values total value for foundation work: | | $800,000 |
| Designer's determination of percent complete (30%): | 0.30 | |
| Total amount due contractor to date: | | $240,000 |
| Total paid to date: | | $150,000 |
| Approved amount of contractor's invoice: | | $90,000 |

---

To perform this review effectively, the designer must have a detailed understanding of the progress of the work. The level of construction inspection required by the wording suggested in Exhibit 4.6 will equip the designer to provide this type of requisition review.

The requirement of a full review by the designer of the contractor's requisition for payment is addressed in Exhibit 4.9.

---

**Exhibit 4.9    Review of contractor's requisition for payment**

3.22  *Review of Requisitions for Payment* Designer shall review each Requisition for Payment ("Requisition") submitted by Contractor pursuant to the Construction Contract and shall recommend to Owner what action to take with respect to each such Requisition. Such review shall consider the mathematical accuracy of the Requisition; the nature and extent of the supporting documentation; the values and percentages shown in the Schedule of Values required to be submitted with the Requisition by the Construction Contract; the amount of the Work performed and the extent to which it complies with the requirements of the Construction Contract; and the updated project schedule reflecting the progress of the Work to date.

*Evaluating change orders and claims*

The owner's ability to bring in a project on time and on budget does not depend entirely on its ability to enter into a construction contract with the desired completion date, and the desired contract price. The owner will have to undertake rigorous change order and claims control in order to maintain schedule and budget objectives. It will be necessary to involve the designer in this process. Therefore, the design agreement should establish the designer's responsibilities with respect to change orders and claims. It should address three issues in this connection.

1   *Base services* The design agreement should clarify that analyzing change orders and claims are part of the designer's base services. There may be some effort to cap the extent to which these services are part of the designer's base services in order to control the designer's base fee.
2   *Focus on entitlement* The design agreement should require the designer to analyze change orders and claims by focusing on entitlement. Entitlement, for these purposes, means the extent to which the construction contract supports the contractor's position.
3   *Written reviews* The designer should be obligated to provide written reviews of all claims and at least the larger change orders. This forces the designer to think through its recommendation carefully and facilitates decision-making by the owner's senior management.

Exhibit 4.10 addresses these objectives.

---

**Exhibit 4.10   Evaluation of change orders and claims**

3.23 *Evaluation of Change Orders* The Designer shall review all change order proposals submitted by Contractor. Such review shall focus on the extent, if any, of Contractor's contractual entitlement to such change order under the Construction Contract, the adequacy of Contractor's documentation, the status of the work that is the subject of the change order proposal, and any other matters relevant to evaluating the change order proposal. Following such review, Designer shall recommend in writing to Owner what action to take with respect to such change order proposal and for what reasons.

3.24 *Evaluation of Claims* The Designer shall analyze all claims submitted by Contractor. Such review shall focus on the extent, if any, of Contractor's contractual entitlement to such claim under the Construction Contract, the adequacy of Contractor's documentation, the status of the work that is the subject of the claim, and any other matters relevant to evaluating the claim. Following such analysis, Designer shall recommend in writing to Owner what action to take with respect to such claim and for what reasons.

---

The designer's frequent and detailed construction inspection, as required by the provision suggested in Exhibit 4.6, will assist it to effectively analyze change orders and claims because of its resulting detailed knowledge of the progress of the work.

### Reimbursable expenses/additional services

For owners, there are two potential problems with reimbursable expenses. First, some designers will attempt to shift reimbursables from a cost center to a profit center. They do this primarily through a mark-up on the direct cost of the reimbursable expenses and by seeking reimbursement for costs which are only partially allocable to the owner's project. Second, there is the possibility that the owner will have to spend an inordinate amount of time confirming the legitimacy of numerous expenditures with small dollar values.

The design agreement should preclude reimbursable expenses except when required and approved in advance by the owner. Exhibit 4.11 illustrates appropriate wording.

---

### Exhibit 4.11   Reimbursable expenses

6.4 *Reimbursable Expenses* The Agreement Sum shall include all expenses of Designer associated with the performance of its services pursuant to this Agreement. Any Amendment that increases the Agreement Sum shall likewise include all expenses associated with the performance of the services that is the subject of the Amendment. Designer shall not seek reimbursement for any expenses associated with performing its services. If Designer will incur expenses not reasonably foreseeable when this Agreement or an Amendment was executed, and such expenses will be incurred solely because of instructions given to Designer by Owner, Designer may seek advance written approval from Owner for reimbursement of such expenses, and if such approval is granted in advance by Owner, Designer may obtain reimbursement from Owner for such expenses.

---

### Ownership and use of documents

Most architects wish to own the design documents which they produce for building projects. The reasons for that include protecting intellectual property and their creative efforts. The architect does not want the owner to be able to use a set of documents for building A for the design and construction of building B with no payment to the architect for the design documents used for building B. The owner should be willing to respect that very strong preference as long as the design agreement allows the owner to use the design documents to the full extent necessary to design, build, and operate the facility. The owner also does not want the architect recreating the exact same building for someone else. Exhibit 4.12 gives an example of two provisions that address all of these considerations.

### Payment provisions

The agreement should establish the total amount the designer will be paid. It should specify the frequency of payments to the designer, the process for requesting payment, and under what circumstances the owner is authorized to withhold payment.

There are two approaches to determining when the designer will get paid. The first is to establish percentages to be paid at the end of each phase of the project. For example, the designer might be entitled to 15 percent upon completion of schematic design; 15 percent upon completion of design development; 25 percent at the completion of construction

---

**Exhibit 4.12   Ownership and use of design documents**

4.1 *Instruments of Service* Designer shall own the plans, specifications, and other documents produced by Designer (collectively, "Instruments of Service") in the course of performing its services pursuant to this Agreement. Designer shall grant to Owner an exclusive license to use the Instruments of Service in connection with the design, construction, expansion, maintenance, and/or reconstruction of the Project. Owner's license shall not extend to, and Owner shall not use the Instruments of Service for any other purpose.

4.2 *Limitation on Use* Designer shall not use the Instruments of Service for any purpose other than to provide its services pursuant to this Agreement without the advance written approval of Owner.

---

documents; 5 percent at the completion of construction bidding and award of the construction contract; and the remaining 40 percent payable at the end of or during construction administration. The advantage of this approach is that it gives the owner control over the satisfactory completion of each phase because the owner doesn't pay for it till the work is satisfactorily completed. However, as projects have increased in size and value, most firms are no longer able to incur payroll and other expenses through the completion of each phase before being paid. For that reason, monthly invoicing has become much more common.

Monthly invoicing presents the owner with the same potential problem as with the contractor. That is, the owner must pay careful attention to ensure it does not pay the designer more in the early stages of the design phase than the value of what the designer has performed or the designer will not give the last phase, construction administration, the attention it needs because of the limited remaining funds.

Another key issue is ensuring that the owner is entitled to whatever documentation it reasonably needs to substantiate the amount requested by the designer. The use of the words "reasonably necessary" or similar wording limits the owner's requests for additional information that has clear and material relevance to evaluating the invoice.

Exhibit 4.13 provides examples of a set of provisions that address the various issues associated with payment of the designer.

---

**Exhibit 4.13   Payment provisions**

6.1 *Agreement Sum* Owner shall pay the sum of $____ ("Agreement Sum") to Designer. The Agreement Sum shall be paid to Designer as provided in Paragraph 6.2.

6.2 *Payment by Monthly Requisition* The Agreement Sum shall be paid to Designer on a monthly basis with the value of the invoice calculated by multiplying the number of hours worked by each person performing the Scope of Services during that month by the designated hourly rate of each such person as set forth in Exhibit C. In no event shall such rates exceed those set forth in Exhibit C except pursuant to an Amendment as authorized by Article 7.

---

6.3 *Invoices* Designer shall submit invoices on or before the fifth (5th) day of the month following the month during which the services that are the subject of the invoice were performed. Owner may request any documentation reasonably necessary to determine if Designer is entitled to the amount set forth in Designer's invoice. Owner shall make payment to Designer within thirty (30) days of the receipt of an acceptable invoice from Designer. Such payment shall be for the full amount of the invoice, unless the amount is disputed by Owner, in which case, Owner shall pay the undisputed amount within such thirty (30) days.

### Schedule for delivery of services

Assuring that the project will be completed by the date on which the owner needs the facility starts with making sure that the design documents are completed by a date which allows the contractor sufficient time to build the project. In order to monitor the timeliness of the progress of the design work, it is recommended to make a schedule for the completion of the design work with appropriate milestones as part of the design agreement. Exhibit 4.14 contains a sample provision which addresses the design schedule.

**Exhibit 4.14   Schedule for the completion of the design documents**

2.3 *Design Schedule* Designer shall perform its services in accordance with the Schedule for Design Services ("Design Schedule") included in this Agreement as Exhibit A. Designer acknowledges that time is of the essence in the performance of this Agreement. Designer further acknowledges that it has reviewed the Design Schedule and that it is a reasonable schedule. The Design Schedule shall only be extended for causes that are beyond Designer's control.

### Deliverables required for management

To maximize the likelihood that the designer will provide its services satisfactorily, it is necessary to manage the designer's performance. That management, in turn, requires that the designer's performance be measured. The most effective mechanism for doing that is a monthly report. The monthly report should describe what work the designer completed the previous month, where the work stands against the design schedule, where it stands against the design budget, and outstanding issues requiring attention.

Exhibit 4.15 provides a sample provision requiring a monthly report by the designer.

**Exhibit 4.15   Requirement for management deliverables**

3.25 *Monthly Reports* Designer shall submit to Owner by the fifth (5th) day of each month a monthly progress report ("Monthly Report") which shall describe the services performed during the preceding month, important issues to be resolved, Designer's progress compared to the Schedule contained in Exhibit A, amounts paid to Designer compared to the Agreement Sum, and what Designer expects to accomplish during the next thirty (30) days. Attached to the Monthly Report shall be the field reports required by Paragraph 3.19 for the preceding month.

*Insurance*

Making sure the designer has an insurance package that provides the owner with the appropriate protections is very important. The required insurance package will depend on the particulars of each project. Whatever the required insurance package, the owner should require the designer to provide evidence that the specified insurance package is in effect at the start of the design engagement and that it remains in effect through the duration of the project. Exhibit 4.16 provides an example of such a provision.

---

**Exhibit 4.16   Requiring insurance to be in effect**

8.2 *Certificates of Insurance* Designer shall, prior to commencing the performance of services on this Project and thereafter as required by Owner, submit to Owner certificates of insurance which substantiate that Designer has insurance policies in force which meet the requirements of Paragraph 8.1. Owner may require that any policy required by Paragraph 8.1 be submitted for its review.

---

*Termination*

Every agreement must address the issue of termination. It is strongly recommended that the owner–designer agreement allow the owner to terminate the agreement for convenience as well as for cause. This is particularly important for public owners who may experience funding issues that make the start or continuation of a project legally impermissible. A termination for convenience provision makes it possible for the owner to terminate the agreement for any reason it deems to affect its convenience which essentially means any reason at all. This type of provision allows the owner to cancel an agreement without having to prove any of the bases for termination typically included in a termination for cause provision.

To be balanced, and acceptable to many design firms, the agreement will allow the designer to terminate the agreement as well. The most usual ground for designer termination is nonpayment by the owner, but failure to abide by the provisions of the agreement in a consistent or material manner is also a common basis.

Exhibit 4.17 contains sample provisions that deal with termination by the designer and by the owner and payment and indemnification aspects of termination by the owner.

---

**Exhibit 4.17   Termination provisions**

10.1 *Designer's Termination* Designer may terminate this Agreement for Owner's failure to make payment in the manner required by Paragraph 6.2 above by providing fifteen (15) days advance written notice to Owner. If Owner fails to make payment or provide a reason why it has not made payment that complies with the requirements of this Agreement within such fifteen (15) days, Designer may terminate this Agreement at the end of such fifteen (15) day period. Designer may terminate this Agreement for any other failure by Owner to comply with the requirements of this Agreement by providing thirty (30) days advance written notice to Owner setting forth Owner's repeated or material failure to comply with the material requirements of this Agreement. If Owner fails to correct such failure or to provide a reason for such failure(s) that complies with the requirements of this Agreement, Designer may terminate this Agreement at the end of the thirty (30) day period.

10.2  *Owner's Termination* Owner may terminate this Agreement for Designer's repeated or material failure to comply with the material requirements of this Agreement by giving Designer seven (7) days advance written notice. If Designer does not correct such failure within such period, Owner may terminate the Agreement at the end of such seven (7) day period. Owner may terminate this Agreement for its convenience by giving Designer fifteen (15) days advance written notice.

10.3  *Payment to Designer* Upon termination, regardless of cause, Designer shall submit within ten (10) days of the effective date of the termination an invoice for the value of the work performed up to the effective date of the termination. If Designer was not terminated for any reason for which it bears responsibility, Designer shall be permitted to seek reimbursement for any reasonable costs solely attributable to such termination. Such costs shall not include anticipated revenues, lost profits, or any other anticipatory or consequential damages of any kind.

10.4  *Indemnification* If Designer is terminated by Owner other than for Owner's convenience, Designer shall hold Owner harmless for all damages and costs arising from Designer's termination, including but not limited to reprocurement costs, loss of use and/or revenue because of delayed completion of the project, attorneys' and other professional services fees, and other costs.

As stated in the Introduction to this chapter, a sample complete owner–designer agreement is included in the Appendix section at the end of the book as Appendix A. That sample agreement contains the provisions discussed in this chapter as well as many others required for a complete agreement.

# 5 Administering the design agreement

## Introduction

Effectively administering a design contract involves three activities. The first is establishing the various baselines against which the owner will monitor the designer's performance. The second is conducting an active monitoring program. The third is taking appropriate measures if the designer is not complying with the requirements of its contract. The chapter starts by discussing each of these activities. Then, the chapter covers additional services. Additional services is designer terminology for change orders. Additional service proposals almost always request additional compensation. If there is a schedule included in the design agreement, and failing to meet the schedule has serious consequences under the design agreement, then additional service proposals may also include requested adjustments to the schedule. This part of the chapter explains how to minimize the number of additional service proposals; how to analyze those that are submitted; and how to value those that have merit. Before getting to these topics, and by way of providing context, the discussion starts with the most common causes of additional services proposals.

The final part of the chapter discusses the designer's standard of care, which is the obligation of the designer to the owner to use care (i.e., avoid errors and omissions) that arise as a result of the designer's performance of its services. It will also address the owner's recourse if these obligations are not met. A key purpose is to focus owners on the obligations owed to them by their designers, and the existence of remedies when those obligations are not met. Most owners expect to have problems with their contractors but not with their designers. In reality, a lot of owners are damaged by negligent mistakes made by their designers. When owners are damaged by designer errors or omissions, they should be no more reluctant to challenge the mistakes of their designers than they are to challenge the mistakes of their contractors.

## Establishing baselines

A baseline is a standard against which the designer's performance can be measured. It may relate to quality of work, amount of work, cost of work, timing of work, or to some other standard. The purpose of the baseline is to establish the expected level of performance for the particular standard so that the owner can determine if the design contract is being performed satisfactorily. Following are descriptions of several recommended baselines.

### Program

The program is the statement of what the project is intended to be. It is usually developed through a study conducted by the owner's own personnel or by a consultant. It is important that the program be stated in as much detail as possible. This enables the designer to develop the clearest understanding possible of what the owner intends the project to be with respect to function, quality, budget, and schedule.

### Deliverables

The term "Deliverables" refers to the items that a designer is contractually required to produce for the owner. The principal deliverables will be the project design documents, which the designer will be required to submit at various stages of design completeness. The designer will be required to submit additional items, and these additional items are also deliverables.

### Schedule

The designer should be contractually required to comply with a schedule. The schedule will indicate when the designer will complete various stages of the design process. Requiring a schedule will not only produce a useful monitoring tool, it will force the designer to think through its approach to completing the work.

### Budget

The amount of the owner–designer agreement constitutes the budget for completing the designer's scope of work. It represents the cost baseline for completing the design services.

## Progress reports

The designer should be contractually required to produce monthly progress reports that discuss the designer's progress. They should describe the designer's progress to date, the major activities undertaken in the reporting period, the major activities projected to be accomplished in the next month, and the key outstanding issues which need resolution.

## Monitoring designer performance

The owner should actively and continuously monitor the designer's progress against the various baselines described in the section on establishing baselines. Various techniques are available for this purpose. The substance of the monitoring effort will be considered first, and then the techniques.

### What to monitor

The owner should review progress against the various baselines.

- *Program* The designs should be reviewed to determine the extent to which they accurately reflect the program.
- *Deliverables* The designs and the other items that the designer is required to submit should match the contractual requirements.
- *Schedule* The amount of design work completed should be compared to the amount called for by the schedule at the time of the review.

- *Budget* The extent of satisfactorily completed work should be compared to the schedule and compared to the amount requested by and paid to the designer. This comparison becomes both more difficult and more important if the design contract provides for monthly billing by the designer, rather than billing on submission of certain deliverables, or on certain percentage completion. In the case of monthly billing, it is critical to confirm that actual progress is being made; otherwise, the owner may shortly find itself having paid a significantly greater percentage of the total fee than the percentage of work completed by the designer.

### How to monitor

There are several monitoring activities that will significantly increase the likelihood of enhanced performance by the designer.

#### Regular meetings

The owner should hold regular meetings with the designer to discuss the progress of the design contract. These meetings may be held monthly, or more frequently if the design work is particularly complicated or sensitive, or if the designer's performance is cause for concern. The meetings should involve a review of the design documents, the budget, the schedule, the progress reports, and any other matters of relevance to the successful completion of the project.

Some of the meetings ought to be held at the designer's office. This will allow the owner to observe first hand who is performing the owner's work, and how the work is progressing.

#### Document discussions and decisions

It is important to have a record of the issues discussed. This provides a record of what issues were considered. It is important to document the decisions reached concerning these issues. In the case of a designer with whom the owner has a contentious relationship, it is very useful to be able to show that a particular concern was voiced on more than one occasion, and to document the designer's response or lack of response to those expressions of concern.

#### Require and utilize reports

It is very difficult to track progress if there are no indicators along the way. That is why a budget, a schedule, and progress reports are important. They allow the owner to have a detailed understanding of how the design contract is progressing. These monthly reports should be required. They should be carefully read, and they should be discussed at the regular meetings, in detail if necessary. These discussions should be carefully documented.

## Designer not complying with requirements of designer agreement

When the designer does not satisfy the requirements of the designer agreement, there are two actions the owner should take.

- *Write letters* The owner should write one or more letters to the designer describing the activities constituting noncompliance. Reference should be made to the specific agreement provisions with which the designer is not complying. (See Chapter Three for a more detailed discussion of effective correspondence.) Writing the letter(s) will cause many designers to correct the deficiencies in their performance.

- *Withhold payment* If the designer does not respond to the owner's letters, the owner should withhold payment for those services that do not meet contractual requirements. Withholding should be preceded by at least one letter that explains in detail how the designer's work fails to comply with contractual requirements and explicitly states that failure to correct the problem(s) will lead to withholding of payment.

## The three gaps

When administering a design contract, there are three gaps to be avoided. These are the vision gap, the approval gap, and the liability gap.

### The vision gap

The vision gap is the difference between the way the owner views the project to be designed and built and the way the designer views it. It is important to close this gap early in the design process in order to save potentially significant costs. These costs can arise in two ways. First, the designer can over-design the project causing the project budget to rise. Second, the designer may recommend or encourage change orders that the owner would not want in order to achieve the designer's view during the construction phase.

### The approval gap

This is the gap between the qualified or limited approval the designer may grant to shop drawings and other submittals and the unqualified approval or rejection which the owner and the contractor require to know the actual contractual status of the particular submittal. This gap can arise if the standard AIA owner–designer contract is not modified because that agreement contains language limiting the purpose of the approval to compliance with the design concept.

### The liability gap

This is the gap between the designer's duty of care to the owner and the owner's potential liability for a defective design. This gap arises because any extra costs incurred by the contractor resulting from a problem with the design documents are costs that the owner will have to pay to the contractor. Whether the owner is able to obtain reimbursement for these costs from the designer will depend on whether the problem with the design documents resulted from the designer not producing the design with the required standard of care. If the problem is not deemed to be a violation of that standard, then the owner will still be liable for the costs to the contractor and the gap will have a specific dollar value.

## Causes of additional service proposals

There are three particularly common causes of additional service proposals from designers. These are changes to the program, schedule slippage, and designer business philosophy. Each is discussed below.

### Changes to the program

Owners will make changes to the program necessitating changes in the design documents. The term design documents as used in this chapter means the plans and specifications included in the construction contract.

These changes arise from four common sources. First, the market for the type of space may change. For example, families downsizing from their large, single family homes to condominiums may be thought to want relatively large kitchens. However, as the design for a particular condominium project progresses, market research may disclose that these families want small kitchens because they don't want to do a lot of cooking. This may cause a need to redesign the unit kitchens.

A second cause of changes in the program is actions by regulatory agencies. For example, the fireproofing for the underside of deck steel may be two hours at the start of design. Subsequently, the fire department, or other agency with jurisdiction, may determine that the appropriate rating is three hours. This would necessitate a change in the design documents.

A third cause is the requirements of the financing sources. Whether a public source of financing, an equity investor, or a debt lender, these sources will invariably have requirements for the projects they fund. If a funding source gets involved in the project after design has begun, a common sequence of events, the design may have to be changed to meet the requirements of the financing source. For example, schematic design may show an office building with a large central atrium, but the equity investor may insist on a smaller atrium to achieve more rentable space. This would require changes in the design documents.

The fourth common reason for changes in the owner's program is changes requested by the ultimate user. This is applicable to the public sector where the design and construction of office buildings, libraries, courthouses, and other public buildings are managed and contracted for by an agency that specializes in these functions but is not the ultimate user. For example, the federal government's principal agency for managing and contracting for design and construction of federal buildings is the General Services Administration, but the ultimate users of these buildings are the other federal agencies and the judiciary. The using agency may determine that its needs have changed since the design process began, requiring changes in the design documents.

A similar situation exists in commercial real estate. The developer of an office building may lease space to a tenant that has different space needs than originally designed. The needs of the tenant may change as its space is being designed. This situation also confronts the real estate departments of operating companies. The needs of a business unit may change as its space is being designed.

All of these changes in the owner's program resulting in necessary changes in the design documents arise from reasons that are not the responsibility of the designer. Therefore, they all represent likely candidates for successful additional service proposals.

### Schedule slippage

If the project takes longer to design than anticipated, a request for additional services is likely. The designer will claim that the additional time, even if there is no additional scope, has raised costs. That claim might have one or both of two potential bases. First, the designer may claim that even though the scope didn't change, the level of effort did. Second, the designer may claim that if the schedule slipped, its costs to perform the same scope increased. This would relate principally to scheduled increases in the salaries of the people producing the design documents.

If the schedule for designing a project slips for reasons that are not the designer's responsibility, it may well have a meritorious claim for additional services. If the extended schedule is related to a necessary increase in the level of effort, the costs associated with the increased level of effort may be recoverable as additional services. If the schedule slips for reasons not the responsibility of the designer, and the work is pushed into the following year and is subject to scheduled salary increases, those costs may well be appropriate for additional services.

If the extended schedule is the designer's responsibility, then any claim for additional services would not have merit. For example, if it took the designer longer to complete the construction documents than called for in the schedule because some of its staff was working on another project longer than anticipated or because completion of the construction documents just took longer than anticipated, the designer would not have a meritorious claim for additional services.

### Designer business philosophy

The third common reason for requests for additional services is designer business philosophy. There are some designers that approach additional services in the same manner that a contractor approaches change orders. That is, the slightest deviation from the program or the slightest slippage in schedule will produce an immediate claim for additional services.

This type of philosophy does not make any specific proposal more or less meritorious. It only increases, sometimes dramatically, the number of proposals to be considered.

## Minimizing additional service proposals

There are five important ways to minimize the number of additional service proposals. These ways are listed in Exhibit 5.1.

---

**Exhibit 5.1    Ways to minimize proposals for additional services**

- Reasonable pricing of the agreement;
- clear drafting of the design agreement;
- owner discipline limiting changes to the design documents;
- proactive management of the design process; and
- rigorous analysis of additional service proposals.

---

### Reasonable pricing

The first step in minimizing the number of additional service requests is to agree on a reasonable value for the services being provided by the designer. The reasonableness of the value of the agreement as a disincentive to additional service requests only applies to lump sum agreements. If the designer believes that it is not being fairly compensated for its work by the payment of the lump sum amount, it has a reason to want to find justification for additional service requests in order to build the agreement sum to a value that is satisfactory to the designer.

If, as a response to this or other concerns, the owner elects to enter into a cost plus agreement, the owner should be fully aware of the risk involved. A cost plus agreement does not usually have a maximum price. The compensation is determined by multiplying the level of effort (i.e., number of hours) of each person assigned to the project times that person's fully loaded rate. The calculation of fully loaded rates is explained in the Valuing additional service proposals section. Unless there is some cap on level of effort or price, the owner's potential exposure for design costs is unlimited. The best way to deal with this risk, if the owner elects to use a cost plus approach to compensation, is to structure the agreement's compensation as a not to exceed value. A contractual provision dealing with a not to exceed value might be worded as shown in Exhibit 5.2.

---

**Exhibit 5.2   Agreement price not to exceed a certain value**

The compensation to be paid by the Owner to the Designer shall be calculated by multiplying the hours worked by each person providing services on the Project times that person's hourly rate as established in Exhibit X. The hourly rates set forth in Exhibit X shall include all of the Designer's direct costs, overhead costs, and fee and shall not be further marked up by the Designer for any reason. The total compensation to be paid by the Owner to the Designer pursuant to this Agreement shall not in any event exceed Five Hundred Thousand Dollars ($500,000) unless authorized in writing by the Owner pursuant to the Additional Services provisions of this Agreement.

---

### Clear drafting

The second step in minimizing the number of additional service proposals is clear drafting of the owner–designer agreement, particularly as it relates to the description of the designer's scope and what constitutes additional services. The first requires a clear description of scope and the second should include foreseeable circumstances that the owner and the designer agree will result in extra compensation.

With respect to scope, most design agreements, even ones that are owner-specific or project-specific have fairly generic language describing the designer's scope of services. The typical agreement divides these services between conceptual design, schematic design, design development, and construction documents, with each stage expected to add a significant level of understanding and detail about the project. It is in the owner's interest to keep the language general because the intent is to provide for a comprehensive set of design documents which can be used by the contractor to build the project. However, for projects where certain portions of the designer's scope is likely to be unusually active or complicated it makes sense to use language that provides more detailed guidance on the designer's basic scope of services. A good example on privately owned projects is the extent to which the designer is expected to participate in obtaining project permits and, particularly, the extent to which redoing designs to obtain such approvals are part of basic services.

The design agreement should obligate the designer to assist the owner in obtaining permits. In many instances, it will be impossible to obtain the permit without some level of assistance from the designer. So the issue becomes at what point is the level of activity required of the designer to obtain a particular permit, or all the required permits, significantly more than could have been anticipated when the design agreement was originally priced. Frequently, this becomes a question of how often the designer has to modify or redo plans.

Perhaps the clearest way to deal with this issue is to have a separate provision dealing with permits. That provision should describe the activities involved in assisting the owner to obtain project approvals. These activities include such things as assisting with preparation of the project description, creating a scale model of the project, assisting with the presentation and/or answering questions at regulatory hearings and community meetings, and preparation of design documents for such presentations. This provision should also address the need to modify the project design documents in response to these hearings and meetings. The designer should expect (i.e., be required by the design agreement) to do some level of redesign related to obtaining project approvals. A sample provision follows in Exhibit 5.3.

---

**Exhibit 5.3    Project approvals**

3.4  *Permits and Approvals* Designer shall assist Owner to obtain all government agency approvals and permits necessary to proceed with the Project. Such assistance shall include preparing applications, drawings, and other required materials; attending community meetings and hearings of public bodies with jurisdiction over the Project; and performing any other activities reasonably required to obtain such permits and approvals. Any of these activities required of Designer for the third and subsequent attempts to obtain the same permit, provided the rejection of the first two applications were not the responsibility of Designer, shall constitute Additional Services for the purposes of Paragraph 7.2.

---

Certain provisions, such as project approvals, need to specify the full scope of the designer's responsibility for that portion of the designer's scope. That is the case even if the owner and the designer agree that portions of the scope will be treated as additional services. The importance of specifying the full scope is so the owner can require the designer to perform all aspects of the scope, regardless of who pays for it.

With respect to the provision addressing additional services, it is useful to enumerate those activities which the owner and the designer can anticipate and which they agree will be additional services. In general, this list will be project specific. An example on private projects is the preparation of marketing materials, which is commonly treated as an additional service. Another common item is delivery of a basic service in excess of the level contemplated as basic service. For example, the design agreement may provide for the designer only analyzing contractor change orders that have design-related issues as base service. If the designer is requested by the owner to evaluate change orders that do not involve design issues that would be additional services.

If there are no specific foreseeable activities that will be additional services, a general but clear description of what constitutes additional services is still very important. Exhibit 5.4 illustrates a sample provision.

---

**Exhibit 5.4    Additional services**

7.2  *Additional Services* Increases or decreases to Designer's Scope of Services shall be made only by Amendment to this Agreement. Services requested of Designer by Owner which are not set forth in Article 2 above shall be considered Additional Services for which Designer shall be entitled to an increase in the Agreement Sum. No increase in the Agreement Sum shall occur as a result of the performance of Additional Services unless Owner has given advance written approval for such Additional Services and such increase in the Agreement Sum.

---

It is vitally important that the design agreement establish that any work required of the designer to correct the designer's errors or omissions is not additional service. Exhibit 5.5 illustrates a sample provision.

---

**Exhibit 5.5   Correction of errors not additional services**

7.4 *Not Eligible as Additional Service* Notwithstanding any other provision of this Agreement, the following shall not constitute Additional Services: preparation of analyses of proposed change orders as required by Paragraph 3.23 until the value of such proposed change orders exceed twenty-five percent (25%) of the original Contract Sum; preparation of analyses of claims as required by Paragraph 3.24; and preparation and administration of the punch list. No service performed by Designer shall be an Additional Service if it is performed because of Designer's negligence, error, or failure to perform services in accordance with the requirements of this Agreement.

---

### Owner discipline

Another important way to minimize additional services is for members of the owner's team to observe two important disciplines. The first is not changing the design documents after they have been approved by the owner, and the second is making sure the owner entity speaks with one voice.

This is particularly applicable to the construction documents (the final stage of the design documents), but it applies to the earlier stages as well. As a purely contractual matter, most design agreements, including the one recommended by this book, and included as Appendix A, require the owner's approval of the design documents at each stage of preparation. Many explicitly state that changes made at the owner's instigation in design documents approved by the owner are additional services. Whether the agreement is explicit or not, ordering modifications to design documents previously approved by the owner will almost always be considered additional services because it will be viewed as the designer having met its obligations and the owner then wanting something else; in other words, an extra.

The second discipline, the owner speaking with one voice, is also very important. The larger the owner's project team, and the more senior the members, the more likely it is the ownership team will have various members who are authorized, or believe themselves authorized, or believe themselves entitled, to speak for the owner. This is a situation to be avoided if at all possible. The most common result of multiple voices is multiple instructions, many of which are not consistent.

It is important to distinguish between two different types of multiple voices. One type is an owner's team on a large project where the project manager, the director of design and construction, or whoever the most senior person authorized to give instructions to designers, consultants, and contractors is too busy to provide all the necessary instructions personally. That person will have a staff of one or more people who may also be authorized to give instructions. In that case, if the owner's organization is properly managed, the staff person or people will be conveying instructions authorized by the senior manager. This situation should provide for reasonably consistent instructions and the efficiency of the senior manager devoting his or her time to the most important issues facing the project.

A second type of multiple voices is where there is more than one senior manager authorized to give instructions to designers, consultants, and contractors. Sometimes a senior manager will not necessarily be specifically authorized but may believe themselves entitled to give such instructions. The key aspect of this second type of multiple voices is that there is no close coordination between the people giving instructions, all of whom appear to the designers, consultants, and contractors to have the authority to give instructions.

This second type of multiple voices is problematic. When it exists, it usually starts at the design stage. The designer receives instructions from more than one team member, and the instructions are, at best, not wholly consistent, and, at worst, inconsistent. This is highly likely to lead to the designer addressing issues with multiple versions of a design document. This, in turn, means either one or more versions will not be utilized or work will have to be redone to conform to the instructions that the owner ultimately chooses as its official instructions. Such work is, in most instances, going to end up being extra work in the eyes of the designer, and probably is extra work under the terms of the design agreement. In short, multiple voices can lead to extra costs for the owner.

This problem, unless cured, then repeats itself in the owner's relationship with its consultants and contractors. The results are the same. The consultants and/or the contractors are likely to end up doing work which will be deemed to be extra work under their respective contracts. Again, this will lead to extra costs for the owner.

A provision in the design, consultant, and construction contracts designating one person as the only person authorized by the owner to give instructions is one way to deal with this problem. However, when a senior member of the owner's team gives a designer, a consultant, or a contractor an instruction, that entity is very unlikely to refuse to carry it out on the grounds that the senior member of the team is not the person designated in the contract to give instructions. The most effective way to deal with this problem is to create a disciplined owner's organization that provides instructions to its designers, consultants, and contractors only through the person authorized to give such instructions.

### Proactive management

The importance of hands-on management of the designer was discussed above. Active management of the designer is also an important mechanism for minimizing additional services.

The regular weekly meetings with the designer provide two important opportunities to mitigate additional services. The first is to address issues at an early stage so they can be resolved as part of basic services. The second opportunity is to head off additional service requests by instructing the designer not to perform services that the owner acknowledges would be additional services.

The owner should be prepared to have additional ad hoc meetings with the designer when necessary to avoid or to challenge additional service requests. The owner assists the designer as well as itself when it can instruct the designer not to undertake certain services before the designer has begun to perform those services.

### Analysis of additional service proposals

The owner should not take additional service requests at face value. They should be reviewed carefully. Those requests that have been the subject of meetings and/or other discussions and have been generally agreed to in advance may not require the same level of scrutiny.

When analyzing a designer's request for additional services, the initial focus should be on entitlement. Entitlement exists if the request has contractual merit; that is, if the design agreement supports the request. The agreement can support the request positively or negatively. It would support the request positively if the request is for an activity that is specifically listed as an additional service in the agreement. The agreement would support the request negatively if the activity that is the subject of the request is not included in the basic services contained in the agreement.

## Valuing additional service proposals

The key point to analyzing the value of additional service proposals is that the owner should not accept the proposed value without performing its own verification. Just like the proposed value of a contractor's change order proposal, the proposed value of a designer's additional services proposal should be carefully reviewed.

Designer proposals for additional services are generally priced one of two ways: lump sum or actual costs. Actual costs, sometimes referred to as "time and materials" (a construction term that doesn't fit design contracts very well) generally means the level of effort (i.e., number of hours) times the hourly rate of each person performing work under the proposal. The hourly rates are usually established in advance in the agreement.

A lump sum proposal is usually prepared by the designer as if it was an actual cost proposal. That is, the designer determines who would perform the services called for by the proposal, what their level of effort is likely to be, and what their fully loaded hourly rates are. Then the level of effort is multiplied by the applicable fully loaded hourly rates. The resulting value may be adjusted up or down if the designer believes there are business reasons why it should ask for more money than actual costs or less money than actual costs.

A proposal based on actual costs, which means the value of the proposal won't be finalized till the work is done, uses the same basic mathematical approach. The value of the additional services will be the level of effort times the applicable fully loaded hourly rate(s). However, under the actual cost approach, there won't be any adjustment of that calculation by the designer.

The owner needs to consider three issues when reviewing the value of an additional service proposal. These are who should take the risk that the work will cost more than originally estimated, is the level of effort reasonable, and are the right hourly rates being used.

The resolution of the issue of who takes the risk of cost overrun determines the pricing approach to be used in the proposal. If the owner wants the designer to take the risk, the owner will request a lump sum proposal. The potential downside to the owner is that the designer will build in contingency for cost overrun, thereby increasing the amount beyond the value that results from multiplying level of effort by the applicable hourly rates. This is an example of a business reason for the designer increasing the lump sum amount beyond the estimated actual cost value.

The owner may prefer to accept the risk that the actual costs will exceed the initial estimate. In that case, the owner will request the proposal to be submitted on an actual cost basis. The downside to the owner is that it needs to actively monitor the level of effort under the additional service amendment to make sure it is reasonable.

The second item the owner needs to focus on is the level of effort. Whether the level of effort is an estimate used to prepare a lump sum proposal or is an actual number based on time sheets, the level of effort is one component of the calculation that produces the proposed or claimed value. Therefore, the owner should assure itself that the designer's level of effort is reasonable.

The second component of the calculation of the value of an additional services proposal is the hourly rate of the person or persons performing services under the proposal. There are two issues to consider when reviewing hourly rates. The first, which should be resolved during the negotiation of the design agreement, is what is included in the rate. Hourly rates as set forth in the design agreement and as used in additional services proposals should be fully loaded.

Hourly rates for designers generally consist of three components. The first is the direct cost, which is the employee's annual salary divided by 2,080 hours (40 hours a week multiplied by 52 weeks). In some definitions of direct cost, related employer costs, such as workman compensation premiums, unemployment premiums, and the like are included in the definition of direct costs. In other definitions, those costs are part of overhead.

The second component of the hourly rate is overhead costs. These costs include, in addition to personnel related costs not included in the direct costs, the costs necessarily incurred in running the business. These include such items as telephones; computers; the salaries of executive, administrative, and marketing personnel (i.e., personnel who do not perform billable work); office rent; and travel. The overhead component of the hourly rate is computed in a three step process. The company's direct and overhead costs are totaled. Then the total overhead costs are divided by the total direct costs, to produce a percentage. Then that percentage is applied to the direct cost component to produce a subtotal hourly rate.

The final component of the hourly rate is profit, sometimes called fee. That component is established either by agreement or internally by the designer. In either case it is a percentage, and the percentage is applied to the subtotal created by applying the overhead rate to the direct cost. That percentage amount is then added to the subtotal to create the fully loaded rate. The process is illustrated in Exhibit 5.6.

---

### Exhibit 5.6    Calculation of fully loaded hourly rate

Assume the proposal involves one person who makes $50,000; that the total of all salary payments to firm employees annually is $1,000,000; that the total of all of the firm's overhead expenses is $1,200,000; and that the established fee percentage is 10%. The person's fully loaded rate is calculated as follows:

| | | | |
|---|---|---|---|
| Direct cost component: | $50,000 divided by 2,080 | = | $24.04 |
| Overhead component: | $1,200,000 divided by $1,000,000 times 100 | = | 120% |
| | 120% of $24.04 | = | $28.85 |
| **Subtotal** | $24.04 (direct rate) plus $28.85 (overhead component) | | $52.89 |
| Fee component (at 10%) | 10% of $52.89 | = | $5.29 |
| **Fully loaded rate** | | | $58.18 |

---

In most cases, the owner will not be involved in the details of the calculation of hourly rates. However, in the case of certain large public projects, the owner may be very much involved in reviewing the rates, and, in general, owners should be aware of how rates are developed so that they can determine with some level of understanding whether proposed rates are reasonable.

## Unilateral amendments

A unilateral amendment is one that is signed only by the owner. This is as opposed to a conventional amendment which is signed by both the owner and the designer.

The unilateral amendment is most useful when the owner and the designer cannot agree on the resolution of a request for additional services. Most typically, this involves a situation where the owner and the designer agree that the designer is entitled to additional compensation, but, notwithstanding extended good faith negotiations, they cannot agree on the value.

The prudent approach to drafting a provision allowing for unilateral amendments is to include in the design agreement what amounts to an appeal mechanism. The agreement could provide that the amendment will be binding on both parties unless the designer provides notice under the dispute or other applicable provision of the agreement. A sample of this approach follows in Exhibit 5.7.

---

**Exhibit 5.7   Unilateral amendments**

7.3 *Unilateral Amendments* Notwithstanding any other provision of the Agreement, Owner may issue a unilateral (i.e., signed only by Owner) Amendment to Designer modifying the Scope of Services and/or any other provision of the Agreement, and Designer shall forthwith perform such Amendment. Such unilateral Amendment shall become binding as written on Designer and Owner unless Designer disputes such unilateral Amendment pursuant to the provisions of Article 11.

---

## Standard of care

The designer has obligations arising from entering into the design agreement. These are contractual obligations. These obligations have been discussed at length in this and previous chapters.

The obligations that are the subject of this section are obligations that arise from the fact that the designer is a professional. Professionals owe their clients an obligation to use care in the delivery of their services. The obligation is commonly referred to as the standard of care. It is the same obligation that the law says lawyers, doctors, and accountants owe their clients. This obligation is established by common law and exists even if there is no design agreement or if there is a design agreement but it does not specifically address this obligation. A design agreement may contractually create a higher standard of care. In that situation, the higher standard becomes applicable to the designer for the project(s) covered by the agreement.

A contractually specified standard of care has two advantages for the owner. First, such a provision clarifies the applicable standard. The provision makes clear whether the standard of care for the particular project is the common law standard or a higher standard.

Second, if a violation of the standard occurs, the owner can allege a breach of contract. This likely would be in addition to a claim of negligence, which is an action in tort. In many states the period during which one can bring a breach of contract action is longer than the period in which one can bring a tort action.

*Basic standard of care*

The obligation of the designer to its client (i.e., the owner), as established by common law, is set forth in Exhibit 5.8.

---

### Exhibit 5.8   Common law standard of care

The designer shall perform its services with the same care as would be used by a reasonable designer performing similar services in the location of the project.

---

This formulation is based on an explicit acknowledgment that the designer's services are not expected to be perfect. A designer (like other professionals, all of whom operate under a similar standard for their respective professions) is expected to perform at a reasonable level of care, but it is not expected to perform flawlessly.

*What constitutes a violation*

The problem for the owner is that, as between the owner and the contractor, anything its designer does that causes the contractor to incur additional costs is the responsibility of the owner, but not everything the designer does that causes these additional costs amounts to a violation of the standard of care. In other words, not every error or omission rises to the level of a violation of the standard of care. So the question becomes what mistakes rise to the level of a violation. An alleged violation must meet the criteria set forth in Exhibit 5.9, each of which is explained further below.

---

### Exhibit 5.9   Criteria for violation of standard of care

1   The design deficiency must be an error or omission; that is, it must be negligent.
2   The error or omission must be significant or part of a pattern.
3   The error or omission must be incapable of correction without cost to the owner.

---

The error or omission must be negligent, meaning it cannot be due to the designer's lack of knowledge. If the designer is unaware of relevant facts that it was the owner's responsibility to provide, then design documents that are deficient because of this lack of knowledge are not negligently deficient. To give some specificity to this point, if the design documents call for a certain level of electrical capacity for an office building, but the owner needs a larger system to run its computer system but doesn't make that clear to the designer, the smaller electrical system, while a design deficiency as against the owner's real needs, is not an error or omission. As another example, if the legal requirements, such as the building code, change during the development of the design documents and the design documents prepared by the designer don't reflect these changes, and are therefore deficient from the building inspector's point of view, the deficiency is not an error or omission unless the design agreement makes it the designer's responsibility to track changes in legal requirements during the development of the design documents.

The error or omission must be significant or be part of a pattern of numerous smaller mistakes. Because professionals are not expected to be perfect, an occasional small mistake simply won't be considered a violation of the standard of care, even if it is an error or omission. For example, in a small office building, with one or two conference rooms, an erroneous specification for the door hardware for that conference room or rooms won't rise to the level of a violation of the standard of care. On the other hand, in a large office building with a lot of private offices, a mistake in the standard door hardware for all those offices may well constitute a violation.

The error or omission must cause the owner to incur extra cost. A mistake can be negligent even if it doesn't cause the owner to incur any extra cost. However, an error or omission, or a pattern of errors and omissions, must lead to extra cost to be considered a violation of the standard of care. There is a common sense aspect to this requirement. If the owner does not incur damages (i.e., costs), there is no violation of any duty owed by the designer to the owner. The duty, again, is not to be perfect but to use reasonable care in the provision of services and thereby to avoid damaging the owner.

What this means as a practical matter is that mistakes made in the schematic design and design development phases are not generally violations of the standard of care. Similarly, mistakes in the construction documents are not violations of standard of care if found during the bidding process. The one exception to these statements is if an error is found at any of these earlier stages and the project is delayed because of required redesign, then any damages resulting from that delay, if significant, may cause the mistake to rise to the level of a violation of the standard of care.

In general for the owner to incur extra costs resulting from a mistake, the mistake must be in the construction documents and must be acted on by the contractor. The contractor must attempt to build it as designed and then discover that it can't be done. Alternatively, the contractor discovers the mistake before the work is performed but precedent work has already been installed which now must be removed in order for the mistake to be corrected. Another example occurs if the work is constructed as designed but it is found to be wrong when the facility is commissioned.

To give further specificity to this discussion, the following is a description of a violation of the standard of care that was included in a settlement of a law suit that alleged a variety of errors and omissions, of which this was one. The cold water pipes for a large commercial building were designed and installed. However, when tested, the system could not withstand the pressure required to move the cold water throughout this very large building. It turned out that the mechanical engineer had not done the appropriate calculations necessary to determine the right size for the cold water pipes. As a result, the pipes had to be removed and replaced at a cost to the owner well in excess of a million dollars.

### Enhancing the standard of care

The standard of care can be contractually enhanced. In other words, the degree care of required of the designer can be increased by agreement. Even under an enhanced standard of care, to be a violation, a mistake must still be negligent and it must still cause the owner to incur extra cost.

Exhibit 5.10 illustrates wording that creates a higher standard of care. This language could be used in the design agreement for any project.

This wording raises the standard of care from a reasonable level of care to the highest level of care. The implication is that a mistake-free or almost mistake-free set of services is required. Under this wording, however, the baseline for determining even the enhanced standard remains practice in the area of the project.

---

**Exhibit 5.10    Language creating higher standard of care**

2.5 *Standard of Care* Designer shall perform the Scope of Services set forth in this
Agreement using the highest level of professional care applicable to the Project.
Such highest level of professional care shall be utilized by the Designer in the
performance of each part of its Scope of Services as set forth in Article 3.

---

Exhibit 5.11 illustrates sample wording that is specifically applicable to large, complicated
projects.

---

**Exhibit 5.11    Language creating enhanced standard of care for large
project**

Designer shall perform the Scope of Services set forth in this Agreement using the
level of professional care of a designer with national experience providing design
services for projects of similar size, scope, and complexity. Such level of professional
care shall be utilized by the Designer in the performance of each part of its Scope of
Services as set forth in the Agreement.

---

This wording raises the standard in two ways. First, the amount of care required to design
a large, complicated project is obviously greater than the amount of care required to
design a small, simple project. Second, this wording dramatically increases the area for
determining the appropriate standard of care from the area of the project to the entire country.
The inquiry changes from is what the designer did considered negligent in the area of the
project to is it considered negligent anywhere in the country.

### Results of requiring enhanced standard of care

The major advantage to the owner of requiring an enhanced standard of care is that it
increases the mistakes that would be considered violations of that standard. That, in turn,
gives the owner access to the designer's professional liability insurance to recover costs that
would not be available under a lower standard because mistakes causing the costs would not
be considered to have been violations of the lower standard of care.

The major disadvantage for the owner is that an increased level of care will probably
result in increased costs to the owner. An enhanced standard of care will lead to increased
premiums for the professional liability insurance carried by the designer because of the
increased risk to the professional liability insurance carrier. In most cases, the design
agreement will provide that any increases in professional liability insurance premiums
attributable solely to that agreement will paid by the owner.

In an extreme case, the enhanced level of care may cause a designer not to be able to
execute the design agreement. That is because the designer may not be able to find any
professional liability insurance carrier willing to undertake the increased risk. When that
occurs, unless the owner knows the design firm very well, it should determine whether it is
considering using a design firm with a history of errors and omissions.

*Owner's recourse if standard is violated*

An owner believing its designer has committed an error or omission must consider the error or omission in the context of the total design services for the project as well in the context of the project's scope and complexity. In the case of a project where the design agreement contains an enhanced standard of care, that standard must be applied when considering the design error(s) and/or omission(s).

If following this analysis, the owner concludes that a violation of the standard of care has occurred it should take action as described in Exhibit 5.12.

---

**Exhibit 5.12   Actions for owner to take when it believes standard of care has been violated**

1   The owner should send a letter alleging one or more violations of the standard of care. The letter should contain a detailed description of each error and/or omission which is alleged to be a violation of the standard of care. The letter should request a meeting to discuss the issues and to develop a plan for solving the problem(s).
2   The owner should meet with the designer to develop a corrective action plan.
3   If the designer is not cooperating, the owner should send another letter which should reference the first letter, reiterate the relevant information of the first letter, and state that continued failure of the designer to cooperate will force the owner to send copies of these letters to the designer's professional liability insurance carrier. The owner may choose to send one more letter without the threat of sending the information to the insurance carrier.
4   If the issue is still not resolved after a reasonable period of time, either because the designer is not responding at all or because it is not negotiating in good faith, the owner should put the designer's professional liability carrier on notice by sending a letter describing the alleged violation of care in detail and attaching the previous letters sent to the designer.
5   If after a reasonable period of time, there is no resolution, the owner will have to commence litigation. It may agree to mediation and/or arbitration before or in lieu of litigation.

---

Proving a violation of the standard of care requires the testimony of an expert witness, usually a designer with extensive experience in the discipline of the designer alleged to have violated its standard of care. The reason this is relevant is to emphasize that violations are not easily proven. An allegation by an owner, however well founded, is still a long way from proof of a violation. The point here is not to discourage consideration of violations—indeed, it is a contention here that owners don't consider this possibility often enough—but rather to emphasize that violations must be carefully considered because of difficulty of proof.

*Indemnification*

Once a violation of the standard of care has been established, the owner's recourse under the owner–designer agreement is to the provisions related to indemnification. These are the provisions which are the basis for the owner seeking to recover the costs associated with the designer's acts or omissions.

Indemnification relates to the extent to which the designer will be responsible for reimbursing the owner for damages caused to the owner by the designer's acts or omissions. Provisions that require indemnification for the result of injury to people or damage to property are most common. It is important that the indemnification obligation also explicitly cover the damages arising from the designer's failure to perform its scope of services as required by the owner–designer agreement. Exhibit 5.13 contains provisions which provide for both types of indemnification.

---

**Exhibit 5.13    Indemnification provisions**

9.1 *Personal Injury or Property Damage* To the fullest extent allowed by law, Designer shall hold Owner harmless and shall reimburse all costs incurred by Owner, including but not limited to the payment of damages, attorneys' and other professional services fees, and other expenses, as a result of personal injury or damage to personal property arising from the negligence, error, acts, or omissions of Designer or any persons or entities for whom Designer is responsible under the Agreement.

9.2 *Failure to Perform According to Agreement* To the fullest extent allowed by law, Designer shall hold Owner harmless and shall reimburse all costs incurred by Owner, including but not limited to the payment of damages, attorneys' and other professional services fees, and other expenses resulting from the negligence, error, acts, omissions, or failure of Designer or any persons or entities for whom Designer is responsible under the Agreement to perform its services in accordance with the requirements of the Agreement.

---

### Ensuring sufficient resources

Winning a dispute with a designer about an error or omission is pointless if there are no funds available to fund the reimbursement of the costs incurred by the owner because of the designer errors. In order to make sure that there are funds available to reimburse those costs, the design agreement should require the designer to carry a specified minimum level of professional liability insurance.

In determining the appropriate minimum, it is important to understand what the professional liability insurance covers. It covers errors and omissions that constitute a violation of the standard of care. It is there to help the designer pay for costs it makes the owner incur as a result of mistakes for which it is responsible.

The answer to how much is the appropriate minimum level of professional liability insurance depends on the size and complexity of the project. The best way to determine the appropriate amount of professional liability insurance to require is to consult the owner's insurance consultant. In general, however, the larger and/or more complicated the project, the more insurance should be required because the more chances there are for designer mistakes. Also, a contractually enhanced standard of care should be accompanied by a relatively larger amount of professional liability insurance. This is because the likelihood of a mistake being a violation of the standard of care, and therefore covered by the professional liability insurance, is increased.

There are two limitations on the upper end of the amount that can be required. First, the designer may be able to obtain the required amount, but the increase in premium over the premium for the designer's standard coverage may be more than either the designer or the owner are willing to pay. Second, at some amount, the designer will not be able to obtain any professional liability insurance coverage because no carrier is willing to write a policy in that amount.

Owners should be aware that very large design firms carry very large professional insurance policies. These firms will not volunteer, and will often refuse, to inform the owner as to the amount of their total professional liability coverage. The owner should not assume that because a large design firm objects to a certain amount of coverage in an agreement, that the firm does not have that much coverage or more. They are trying to keep from putting their total policy at risk in any given project. This may not be unreasonable, but the point from the owner's perspective is that requiring a large design firm to come up with a reasonable amount of insurance on a large, complicated project may not actually represent a significant economic burden for that firm; it may only be something they are trying to avoid for business reasons.

Owners should also require their designers to include in their subconsultant agreements a requirement that the subconsultants have professional liability insurance at the same level as the designer or at some specified level that reflects the discipline and therefore the risk of error of each subconsultant. Even though it is the prime designer and the prime designer's policy that will initially respond to any claim by the owner for a subconsultant's mistake, it is important to have this requirement so that there are funds available to replenish the designer's liability policy, thus preserving resources that would be potentially available to the owner in case of further errors or omissions.

## Limitation on damages payable by the designer

The main types of damages that are potentially recoverable for errors and omissions are direct and consequential. Direct damages are those damages that result directly from the covered act or omission. They are the costs of fixing the problem. In the design and construction context, this generally means those costs related to redesign, removal and replacement, or modification, of a portion of the construction work. Consequential damages, sometimes called special or indirect damages, are those damages that result from the act or omission but are in addition to and unrelated to the costs of actually fixing the problem. This usually means such things as the extra debt service paid if the project duration is extended, the lower value of the rental income or the unit sales prices if these values have dropped while the problem is being fixed, and other similar, indirect, but nonetheless real, consequences of the problem.

Consequential damages should be a reimbursable cost to the extent they are actually incurred and directly attributable to the covered act or omission. There are two reasons consequential damages should be payable. The first is that they are real costs (if not, they should not be recoverable) incurred by the owner, which the owner would not have incurred but for the designer's mistake. Second, allowing recovery of consequential damages provides a more significant potential recovery in case of a mistake and, therefore, a stronger incentive to avoid mistakes.

Designers will frequently seek to limit their damages to direct damages. That is because they are concerned about being held financially responsible for damages they did not create. Two forms of compromise to allay this fear make sense. The first is already suggested above by making clear in the indemnification provision that consequential damages are payable only to the extent that they result directly from the designer's mistake. The second potential compromise is to limit the amount of consequential damages that are payable under this provision. The compromise amount should be large enough to cover a significant amount of damage and serve the intended incentive purpose while being capped at a number that provides a sense of proportion to the designer.

The issue of limiting damages refers to the amount payable if there is a violation of the standard of care. The preferred position for owners is no limitation. This allows for full reimbursement for the costs associated with any violation. The examples in Exhibit 5.12 follow this recommendation and provide for indemnification with no limitation (except as may be in effect due to applicable state law).

Designers will resist this position on two grounds. The first is that the fee they are receiving is much smaller than their potential liability, if the liability is unlimited. Their second basis of resistance to unlimited liability is that the only assets they have are the personal property of their firm so a recovery against the firm beyond the value of those assets will put the firm out of business, which is not an appropriate result from a single mistake. On that basis, designers will seek to limit their liability to the amount of their fee. Their fall-back position is that their damages should be limited to the amount of their available professional liability insurance. (It should be remembered that the designer's professional liability insurance covers all of its work, so in the absence of a project endorsement, the amount available for the owner's project may not be the face amount of the policy.)

There are two levels of acceptable compromise. The first and more acceptable is that the designer's damages are capped for all types of damages (a cap on just consequential damages was discussed above), provided the cap is a significant amount (as discussed above, one that is large enough to cover a significant amount of potential damages while providing some sense of proportion to the designer). The cap should be well above the combined amount of the professional liability policy and the designer's fee. The second compromise is to limit the indemnification recovery to the combined value of the available professional liability insurance and the fee. This is particularly appropriate for smaller projects and/or projects that are physically simple and therefore don't represent significant risk of designer error or where the likely results of such errors are not likely to be financially significant. In no case should the owner agree to a cap consisting of just the fee, because all designers have professional liability insurance and that insurance should be available to the owner; nor should the owner agree to a limitation of liability consisting of just the amount of available professional liability insurance because that would mean the designer would not suffer any direct financial consequences of its mistake.

# 6 Drafting and administering consultant agreements

## Introduction

In addition to retaining a design team, the owner will often hire one or more consultants directly (i.e., the relationship is direct between the owner and the consultant as opposed to the consultant being part of the designer's team). This is particularly true on larger projects.

This chapter addresses the issue of which entities should be on the design team and which entities should be consultants hired directly by the owner. It also discusses the differences between design agreements and consultant agreements, and how those differences should be reflected in the drafting of the consultant agreement. The chapter concludes with a section on administering consultant agreements.

A sample consultant agreement appears in the Appendix section at the end of the book as Appendix B.

## Member of design team or consultant

For the purposes of this chapter, being a member of the design team means being a subconsultant to the lead design firm. One way to determine whether an entity should be a member of the design team or a consultant direct to the owner relates to the extent of its involvement in design. If the entity contributes to design in any way it should be on the design team; whereas if the entity provides information on which design is based, the entity will be a consultant. This should be seen as a guideline aimed at helping make the decision, but it is not a legal requirement or even a hard and fast rule. A couple of examples will illustrate the choice. A surveyor assembles and confirms information on which design is based, so it may well be a consultant. The structural engineer participates in developing the design, so it would be on the design team.

Certain disciplines require a little more attention to determine the appropriate contractual status. For example, a geotechnical firm, if one is involved, is developing information when it is doing soils investigation, but it is doing design work when it is working with the structural engineer to develop a design that responds to the anticipated lateral loads.

Another such example is a curtainwall firm. If the firm is assisting the architect to develop the curtainwall design, it should be a member of the design team. If on the other hand, it is assembling wind force or other information that will be used in the design, it may serve as a consultant to the owner.

In the end, the determination of whether a firm will be on the design team or be a consultant will be made by the owner to best meet the owner's interests. One factor may be that for some projects, the services of one or more consultants may be needed before the designer is selected. Another consideration is keeping potential liability issues as simple as possible.

This would argue in favor of making as many firms and disciplines subconsultants to the lead design firm as possible.

## Differences in consultant and design agreements

There are three major differences between consultant agreements and design agreements. Exhibit 6.1 lists these differences.

---

**Exhibit 6.1   Major differences between design and consultant agreements**

The major differences include:

- scope of services;
- ownership of work product; and
- role during construction administration.

---

### Scope of services

The consultant will not be performing design services, so the schematic design, design development, and construction documents phases, and the usual breakdown of services for each of those phases, will not usually apply to the consultant's scope of services. The scope of services will have to be prepared on a project-specific, consultant-specific basis. In some cases the design phases will be relevant to a consultant's scope of services, but in most cases they won't be relevant.

The key drafting issues concerning consultants' scope of services are clarity and comprehensiveness. The scope of services should be described clearly and comprehensively for two reasons. First, it is important to make sure the consultant provides all needed services. Second, a clear and comprehensive scope description will reduce, or even eliminate, requests for additional services by the consultant.

### Ownership of work product

Consultants generally do not expect to own the deliverables, or the analyses which underlie those deliverables, that they produce for their clients. This is because they typically issue reports describing the process for, and the results of, their data gathering. This is not a creative process, and, in most cases, does not give rise to a copyright. As a result, the consultant agreement will provide for the owner to obtain title to the consultant's work product. Designers, on the other hand, engage in a creative process which can lead to a copyrightable work product. For that reason, among others, many designers expect to retain ownership of the design documents. The design agreement will give the owner a license to use the design documents.

### Role during construction administration

Most consultants will not be involved in construction administration. Therefore, the various provisions describing involvement in construction administration that are typical of a design agreement will not appear in a consultant's agreement.

There are exceptions to this general point. In those cases, it is important that the construction administration portion of the consultant's scope of services be described in detail for two reasons. First, it is important to document the owner's and the consultant's agreement on exactly what services the consultant will be providing during construction administration. Second, it is important to make sure there is no overlap or duplication between the consultant's construction administration services and the designer's construction administration services.

## Similarities between consultant agreements and design agreements

In most other respects, the consultant agreement should be similar to the design agreement. Exhibit 6.2 lists the most important provisions.

---

**Exhibit 6.2    Key provisions of consultant agreements**

Key provisions of consultant agreements include:

- standard of care;
- payment;
- additional services;
- insurance;
- indemnification; and
- termination.

---

These provisions of the consultant agreement should be drafted in the same manner as in a design agreement. The key aspects of these types of provisions are briefly reviewed below in the order in which they appear in the sample consultant agreement in Appendix B.

When drafting a consultant agreement, the reader may wish to review Chapter Four on drafting design agreements as well as the complete sample consultant agreement in Appendix B.

### Standard of care

In general, consultants assisting owners in the design and construction of their facilities are providing professional services. As such the consultants owe their clients the duty to provide their services with care (i.e., without negligence).

The standard of care is the appropriate equivalent of the standard applicable to designers. The standard for consultants is stated in Exhibit 6.3.

---

**Exhibit 6.3    Standard of care for consultants**

3.5 *Standard of Care* Consultant shall perform its services with the same care as would be used by a reasonable consultant performing similar services in the location of the project.

---

As with the standard for designers, the standard for consultants can be contractually enhanced. And enhancing a consultant's standard of care can create the same potential problems as increasing the standard for a designer: increased premiums for professional liability, or inability to get such insurance because of the higher standard. However, it is the rare consulting service that is so critical to a project that it justifies a higher standard of care.

### Payment

The agreement should establish clearly whether compensation is to be on a cost plus basis, a lump sum basis, or a not to exceed basis. If it is to be either a lump sum or a not to exceed agreement, the payment provision should include a statement that in no event will the total payments to the consultant exceed the lump sum or not to exceed value except as that value may be increased by amendments to the consultant agreement.

The payment provisions should include a description of how invoices will be prepared and calculated. For cost plus and not to exceed agreements, invoices will usually be prepared by multiplying the fully loaded rates of each person that performed services during the billing period, generally monthly, times the number of hours they worked on the project. For lump sum agreements, the invoice may be prepared in the same manner, or it may be prepared on the basis of percentage completion of the consultant's scope of work.

Another important provision in the payment section is the one dealing with documentation. The agreement should provide that the owner may require any documentation that it reasonably believes is necessary to substantiate the amount being requested in any invoice.

### Additional services

It is important that consultant agreements contemplate two types of amendments. The first is the conventional amendment executed by both parties. The second is the unilateral amendment, which is signed only by the owner and becomes effective unless disputed in a timely manner by the consultant.

The second important issue relating to additional services is the consultant's burden of proof. The agreement should state that the burden is on the consultant to prove any request for additional service has merit and is in the appropriate value.

The third important issue relates to documentation. The agreement should provide that any request for additional services must be accompanied by sufficient documentation to reasonably substantiate the merit and value of the request. The agreement should further provide that the owner has the right to request any additional documentation reasonably necessary to determine if the request has merit and/or is in the appropriate amount.

### Insurance

Because of their standard of care and their indemnification obligation, it is important to require consultants to carry professional liability insurance. The amount will depend on the consultant's scope and the size and complexity of the project. In general, consultants are required to carry lower amounts of professional liability insurance than designers.

The agreement should give the owner the right to review and approve the policies required by the agreement. A certificate evidencing all the required coverages should be required before the consultant starts to provide any of its services. Finally, the agreement should provide that none of the policies can be cancelled or their coverages reduced below the levels required by the agreement without 30 days notice to the owner.

### *Indemnification*

The agreement should require the consultant to indemnify the owner for damages caused by the consultant through its negligence. From the owner's standpoint, the best provision is one that requires the consultant to indemnify the owner for any damages arising from the provision of its services. However, consultants, like designers, are concerned about limiting the extent of their indemnification obligation. The reasonable compromise for consultants is the same as that for designers; namely, limiting the extent of the liability for damages to the amount of their professional liability insurance plus their fee.

### *Termination*

The consultant agreement should provide for two types of termination. There should be a provision describing termination for cause (i.e., not performing as required by the agreement) and one describing termination for convenience (i.e., reasons solely attributable to the owner's needs).

The provision dealing with termination for cause should state that no further payments will be made to the consultant following termination. Should the cost of completion of the consultant's scope of services exceed the unpaid balance of the agreement sum, the consultant shall be responsible for the difference; and if the cost of completion is less than the unpaid balance, the difference shall be paid to the consultant upon completion of the project.

The provision addressing termination for convenience should provide that the consultant will be compensated for the total value of all services provided up to the date of the termination minus the value of all previous payments. If the owner wishes to reimburse any of the consultant's costs associated with a termination for convenience, the obligation to reimburse should be limited to reasonable direct costs solely caused by the termination for convenience and which the consultant could not reasonably avoid. However, that provision should also state that the owner is not obligated to pay for any loss of anticipated revenue, lost profits, or other consequential damages of any kind.

## Administration of consultant agreements

A consultant agreement should be administered in the same manner as design agreements. Just like design agreements, the key to successfully administering consultant agreements includes frequent meetings with the consultant, hands-on involvement with the performance of the scope of services, and a rigorous review of any proposed additional services.

# 7 Key issues in drafting the construction contract

## Introduction

When drafting a construction contract, it is important to consider two issues: coordination with the design agreement and risk allocation. After discussing these two issues, this chapter will identify a number of specific issues in drafting the construction contract that can significantly impact the successful completion of the project, and will suggest possible wording to address these issues.

A complete sample construction contract is included in the Appendix as Appendix C. This contract assumes a standard design-bid-build delivery method. It also assumes the project in question is a relatively standard project in terms of scope, size, and complexity. There is no specific size or scope for which this contract is inappropriate. However, for projects of over $20 million, the owner should consider whether additional or more detailed provisions may be in order.

The construction manager (CM) at risk delivery is gaining popularity among public and private owners. Therefore, certain contractual provisions unique to the CM at risk delivery method are discussed later in the chapter. All the provisions discussed earlier in this chapter should also be included in any CM at risk contract.

## Coordination with design agreement

It is necessary that the design agreement and the construction contract be coordinated. What this means is that the wording establishing certain obligations should be the same in each document. For example, the provision in the design agreement describing the designer's responsibilities for construction inspection should be repeated in the construction contract. This is because the contractor needs to understand how construction inspection will be conducted and what its responsibilities in connection with this activity will be. The same is true for schedule administration, review of requisitions for payment, and a variety of other issues.

The need for this coordination between the design agreement and the construction contract is another reason why an owner who does any appreciable amount of construction should have its own standard documents. Using the AIA designer agreement, the design agreement of choice for many designers, with a construction contract drafted by the owner's attorney creates a significant potential for conflicts between the documents.

## Risk allocation

When drafting a construction contract, the owner should carefully consider the issue of risk allocation. Risk allocation refers to which party bears the responsibility for the costs associated with building the project. In a lump sum transaction, the contractor assumes the risk that it can build the project for the agreed upon price. The owner's risk is that the price will be within its budget. What is more complicated is determining which party will bear the cost of unforeseen events such as differing site conditions, unusually severe weather, and correction of defective work, just to name a few.

These types of events have direct costs associated with their resolution. They may cause delay and there may be additional costs associated with the delay. For example, while excavating a hotel foundation, the contractor may encounter a ledge not shown on, or referred to, in the contract documents, and not known to exist at the site. Removing the ledge has a cost. If removing the ledge causes a delay in the project schedule, the delay may cause the owner to have to pay additional borrowing costs and suffer a loss of revenue because the hotel did not open when planned. Risk allocation involves allocating responsibility for these and other types of potential costs.

From the owner's standpoint, risk allocation involves balancing possible present costs against potential future costs. The provision addressing differing site conditions is a good example. A construction contract with no differing site condition provision, or with a provision saying that the contractor owns the conditions at the site as encountered, will cause the contractor to include a significant contingency in its bid for the costs associated with unknown conditions. However, the contractor will have no entitlement to any additional costs associated with any differing site conditions it encounters in performing the work.

A provision stating that the contractor is entitled to all extra costs associated with differing site conditions will cause the contractor to believe a contingency specifically for differing site conditions is not necessary because costs will be reimbursed as encountered. This will lead to a lower bid. If the contractor encounters significant differing site conditions, it may be entitled to a large increase in the contract price.

The owner may wish to share the risk, as opposed to shouldering all of it or assigning all of it to the contractor. Such a risk sharing provision might state, in the case of a contract involving excavation, that the contractor will be entitled to recover extra costs for unknown conditions encountered below 10 feet (or some other depth).

When drafting its construction contract, the owner should always be sensitive to the issue of risk allocation. The owner will want to consider whether it rather take the risk that it will receive a bid containing a contingency to cover the potential costs associated with a particular unforeseen event or circumstance; or whether it rather take the risk of having to pay larger change order costs associated with that event or circumstance; or whether it should seek to balance the risk as between itself and the contractor.

## Specific issues to consider when drafting a construction contract

There are a number of issues that should be carefully considered when drafting a construction contract. The following issues deserve particular attention.

- *Contractor responsible for acts and omissions of its subcontractors* This provision makes the prime contractor responsible for all the acts and omissions of its subcontractors. This provision has become even more important because prime contractors, whether general contractors or construction managers, are no longer self performing any significant amount of work. Virtually all of the work involved in building a project is being performed by trade contractors who, from a contractual standpoint, are all subcontractors.

- *Schedule* It is important that the contractor be responsible for completing the project when the owner needs it completed, and that the construction schedule be a key mechanism for facilitating the achievement of that objective.
- *Indemnification* Most construction contracts require the contractor to indemnify the owner for damages associated with bodily injury and property damage. The owner should also be protected from the consequences of the contractor not performing the work as required by the contract.
- *As-built design documents* The owner needs a set of drawings and specifications that show how the project was actually built, as opposed to how it was designed. They will be important to the owner personnel that will operate and maintain the facility. They may also become important for claims defense purposes.
- *Safety* The owner needs the contractor to be focused on safety for moral reasons (concern for workers' welfare) and for cost control reasons (delays and/or premium increases related to injuries or death can be expensive). This provision should discuss how the contractor will address its responsibility for safety.
- *Quality control* It is important to make the contractor's responsibility for quality control explicit. This provision should address how the contractor will be required to implement this responsibility.
- *Withholding payment* The owner should be able to withhold payment to the contractor if the contractor does not perform the work as required by the contract. Common law principals of contract law may allow withholding on that basis; however, the owner has more protection if the construction contract makes that basis for withholding payment explicit.
- *Termination* Among other reasons, the owner must be able to terminate the contractor for failing to perform the work as required. It is also advisable that the owner be able to terminate the contractor for convenience.
- *Owner can obtain documentation* The owner will want the ability to assure itself of the contractor's entitlement to any position the contractor takes. Certain provisions relating to specific contractor requests—such as those relating to requisitions for payment, change orders, and claims, include a requirement for documentation. However, the owner should have the ability to obtain relevant documentation related to other issues as well.
- *Waiver* As discussed in Chapter Three, owners must be alert to avoid actions or failures to act that may have the effect of waiving contractual requirements. The construction contract should assist this effort by including a provision that seeks to minimize the acts or failure to act which actually constitute a waiver, as well as minimizing the effect of those that are deemed to be waivers.

### Contractor responsible for acts and omissions of its subcontractors

It is important to explicitly establish the contractor's responsibility for the acts and omissions (i.e., the mistakes) of its subcontractors. That responsibility is a significant component of the value that the contractor provides. This provision should explicitly assign the risk of subcontractor mistakes to the contractor. Exhibit 7.1 provides an example.

---

**Exhibit 7.1   Contractor responsible for acts and omissions of subcontractors**

3.12 *Responsible for Subcontractors' Acts and Omissions* Contractor shall be fully responsible for all the acts and omissions of all its Subcontractors at whatever tier.

---

The definition of subcontractor in the sample contract in Appendix C is all inclusive. Paragraph 3.4 states in part:

> Any organization, person, or entity furnishing labor, services, materials, and/or equipment to Contractor shall be a Subcontractor for the purposes of this Contract, regardless of whether it has a Subcontract with Contractor or another Subcontractor, and regardless of whether it provides labor or services or only provides materials or equipment.

This definition eliminates the need to decide in what category an entity fits. For example, it means it is not necessary to decide if an equipment manufacturer that participates in installation and/or training on use of the equipment is a vendor or a subcontractor.

### Schedule

The contractor must be responsible for completing the project when the owner needs it completed. To facilitate that objective, the contract should require the contractor to prepare, submit, and update a construction schedule.

The contract should specify what type of schedule is required. For most projects of any size, the contract should require a critical path method schedule. That is because breaking the project into a more detailed list of activities, focusing on logic and durations, and tracking the critical path will assist the contractor to focus on how to complete the project to meet the owner's schedule needs. On smaller, more routine projects, a bar chart schedule may be sufficient.

Regardless of the type of schedule required, the submission of the initial schedule and regular updates, and the owner's approval of those submissions, is sufficiently important that they should be made a precondition to the owner making any progress payments. The requirement for submission of the initial schedule should be a precondition to the contractor receiving its first progress payment. Similarly, the owner should be able to consider any subsequent requisition for payment incomplete if it does not include a schedule update which the owner can approve.

Updating the schedule is important. The contractor should be required to submit schedule updates for approval on a monthly basis. The updates should be required to reflect actual progress in the field. All schedule submissions should be required to show the project being completed as of the contractually specified completion date. That is the date originally established plus any time extensions granted through change orders. The contract should require the contractor to regain schedule. When the contractor falls behind the schedule for reasons for which it is responsible, the contractor should be required to get back on schedule at its own cost.

Lastly, the nature of the schedule approval should be clarified. The contract should state that approval of the contractor's schedule is limited to an assurance that the schedule is realistic. The approval does not deal with whether the methods and sequences indicated by the schedule are the best approach to building the project, only whether they offer a realistic plan for completing the project by the contractually required completion date.

Exhibit 7.2 is an example of a schedule provision that reflects the above considerations.

### Indemnification

The owner should be protected from the consequences of the contractor not performing the work as required by the contract. Many indemnification provisions address only damages

---

**Exhibit 7.2  Schedule provision**

3.13 *Contractor's Schedule* Contractor shall prepare, submit, and update a project schedule in accordance with the requirements of this Paragraph 3.13. Within fifteen (15) days following execution of this Contract, or by a later date if approved in writing by Owner, Contractor shall submit an initial schedule for approval by Owner. The submission and approval of such initial schedule shall be a precondition to Owner making any payment pursuant to this Contract. Thereafter, Contractor shall submit an update of such schedule monthly for approval as part of its requisition for payment until final completion, or until Owner relieves Contractor of this responsibility in writing. Failure to submit a schedule update by Contractor, or to obtain Owner's approval for such update, may serve as a basis for Owner to deem the associated requisition for payment incomplete and not ready for processing until accompanied by such approved schedule update. Unless another format is approved in writing by Owner, all schedules required by this Paragraph shall be critical path method ("CPM") schedules. If Contractor falls behind the most recently approved schedule for reasons that are its responsibility, Contractor shall submit a recovery schedule for approval by Owner and shall implement such recovery schedule at no additional cost to Owner. A written instruction by Owner to Contractor directing it to submit and implement such recovery schedule shall not constitute an instruction to accelerate. Approvals required by this Paragraph 3.13 shall be limited to determining that any schedule submission provides a realistic approach to completing the Project within the Contract Time and that any update also accurately reflects the progress of the Work.

---

resulting from bodily injury or property damage. It is important that the owner also be protected from damages resulting from the work not being performed as required by the contract. Exhibit 7.3 contains a provision that addresses this concern.

---

**Exhibit 7.3  Indemnification**

3.16 *Indemnification* To the greatest extent permitted by law, Contractor shall hold Owner harmless and shall pay all damages and all reasonable costs, including but not limited to attorney, consultant, and other professional service fees incurred by Owner to the extent such damages and costs are paid or incurred by Owner as a result of Contractor's failure to perform the Work as required by the Contract Documents or as a result of bodily injury and/or property damage caused by Contractor's negligence. Contractor's responsibilities under this Paragraph shall include all acts and omissions of its Subcontractors.

---

### As-built design documents

The owner needs a set of design documents that show how the project was actually built, as opposed to how it was designed. They will be important to the owner personnel that will operate and maintain the facility. They may also become important for claims defense purposes.

This type of provision typically talks about drawings. However, the specifications will also be important for the same reasons. Therefore, this type of provision should refer to as-built design documents, as opposed to as-built drawings. Exhibit 7.4 illustrates an appropriate provision.

---

**Exhibit 7.4   As-built design documents**

3.17 *As-Built Design Documents* Contractor shall maintain a fully conformed set of the Design Documents for the Project which shall be available for Owner's and Designer's inspection at any time. Upon completion of the Work, Contractor shall submit to Designer for Designer's approval a complete set of the as-built Design Documents.

---

*Safety*

The owner needs the contractor to be focused on safety for moral reasons (concern for workers' welfare) and for cost control reasons (delays and/or premium increases related to injuries or death can be expensive). Exhibit 7.5 illustrates a provision that discusses how the contractor will address its responsibility for safety.

---

**Exhibit 7.5   Safety**

3.18 *Safety* Contractor shall be responsible for all aspects of safety on the project. It shall be Contractor's responsibility to avoid, to the maximum extent feasible, bodily injuries and property damage resulting from performance of the Work. At least thirty (30) days prior to commencing work at the Site, Contractor shall submit to Owner for Owner's approval, a safety plan indicating how Contractor shall maximize safety on the Project, and Contractor shall not commence work at the Site until such safety plan is approved in writing by Owner. Owner's approval for the purposes of this Paragraph 3.18 shall be limited to determining that Contractor's safety plan represents a reasonable approach to ensuring safety on the project. Contractor shall designate a key member of its project team as Safety Manager. The Safety Manager shall be responsible for implementing Contractor's safety plan.

---

*Quality control*

It is important to make the contractor's responsibility for quality control explicit. Quality control is the process by which the contractor ensures that the actual construction work complies with the requirements of the contract documents. It is important to remember that the standard of quality that the contractor is required to meet is the standard established by the contract documents. It is not the owner's, the designer's, or anyone else's abstract notion of quality.

The term "quality control" is often associated with the term "quality assurance." Quality assurance is the owner's function. It is the process of determining that the contractor is properly implementing its quality control responsibilities.

The provision of the contract dealing with quality control should address how the contractor will be required to implement this responsibility. While requiring the contractor to submit a quality control plan on any project makes sense, the complexity and level of effort required of the plan should be reasonably related to the size and complexity of the project. Exhibit 7.6 contains an example of a quality control provision.

---

### Exhibit 7.6    Quality control

3.19  *Quality Control* Prior to commencement of any work, Contractor shall submit to Owner for Owner's approval a quality control plan indicating how Contractor intends to ensure that the Work complies with the requirements of the Contract Documents. Contractor shall not commence work at the Site until such quality control plan has been approved in writing by Owner. Contractor's quality control plan shall include the identification of a member of Contractor's staff who shall be responsible for the implementation of the plan. Owner's approval for the purposes of this Paragraph 3.19 shall be limited to a determination that the plan represents a reasonable approach to ensuring the Work complies with the Contract Documents.

---

### *Withholding payment*

The owner should be able to withhold payment to the contractor if the contractor does not perform the work as required by the contract. Common law principals of contract law may allow withholding on that basis; however, the owner has more protection if the construction contract makes that basis for withholding payment explicit. Exhibit 7.7 illustrates such a provision.

---

### Exhibit 7.7    Withholding payment

6.6  *Withholding Payment* Owner may withhold all or a portion of a progress payment to Contractor if Owner reasonably believes that Contractor has failed in one or more significant ways to perform the Work as required by the Contract Documents; has failed to make timely payments to Subcontractors as required by Paragraph 6.7 below; or has breached this Contract in any other material manner.

---

### *Termination*

The owner must be able to terminate the contractor for failing to perform the work as required by the contract documents. It is also highly advisable that the owner be able to terminate the contractor for convenience. There are three key issues.

- *Grounds for termination* The contract should, in addition to the usual grounds related to the contractor's financial condition, establish failure to perform the work as required by the contract as a ground for termination.

- *Process of termination* It is important that the process of termination be as simple as possible.
- *Payment following termination* It is important to preclude payment of any money not associated with the performance of work, except for reasonable direct costs solely attributable to the early termination when the contractor is terminated for convenience.

Exhibit 7.8 contains provisions that address these issues.

---

**Exhibit 7.8   Termination**

12.1 *Termination for Cause* Owner may terminate the contract if contractor:

    a. fails materially or persistently to perform the Work as required by the Contract Documents;

    b. refuses or otherwise fails materially or persistently to respond to instructions issued by Owner, Designer, or any other person or entity authorized by this Contract to issue instructions on Owner's behalf;

    c. refuses or is unable to commit adequate levels of labor, materials, and equipment to perform the Work within the Contract Time or is otherwise unable to complete the Work within the Contract Time for reasons for which it is responsible; or

    d. is unable to provide evidence of financial ability to complete the Work when such evidence is requested by Owner in response to a bankruptcy filing by Contractor or other evidence from which Owner may reasonably conclude that Contractor may be experiencing significant financial problems.

12.2 *Process of Termination* Owner shall, on the basis of one or more of the grounds for termination enumerated in Paragraph 12.1, send Contractor a certified letter, return receipt requested, or by any other method that documents receipt by Contractor, stating in detail each of Owner's grounds for termination of Contractor and giving Contractor seven (7) days to cure each such ground for termination. If at the end of the seven (7) days, Contractor has not cured each such ground for termination, or for those grounds which cannot be fully cured in seven (7) days, made reasonable progress in curing such grounds, Owner shall terminate Contractor by delivering in hand to Contractor at Contractor's office at the Site written notice that Owner has terminated the Contract pursuant to its letter described in this Paragraph ("Termination Notice"). Contractor shall immediately cease performing the Work, and Owner shall immediately thereafter take possession of the Site, including any of Contractor's materials and/or equipment, and shall complete the Project in whatever manner it deems most appropriate. Copies of all correspondence required by this Paragraph shall be sent to Contractor at the addressed specified by Paragraph 15.7 and to Contractor's surety.

12.3 *No Further Payment to Contractor* Effective as of the date of termination as established by the Termination Notice, Owner shall make no further payments to Contractor until the Work is finally complete. Upon final completion, Owner shall determine the cost to complete the Project ("Completion Cost"), which shall include the cost of completing work not performed by Contractor, the cost of completing any work started but not completed by Contractor, and the cost of correcting any defective work performed by Contractor. If the Completion Cost of the Project exceeds the balance of the Contract Price unpaid to Contractor at the time of termination, Contractor shall pay the difference to Owner. If the Completion Cost of the Project is less than the balance unpaid to Contractor at the time of termination, Owner shall pay the difference to Contractor.

12.4 *Termination for Convenience* Owner may terminate this Contract for its convenience at any time by giving Contractor at least seven (7) days written notice by certified mail, return receipt requested or any other manner which documents receipt by Contractor. Such notice shall include instructions to Contractor concerning how to implement the termination. Upon receipt of such notice, Contractor shall faithfully perform the instructions set forth in such notice. Final payment to the Contractor following termination for convenience, which shall be requested and paid in accordance with the provisions of Article 6, shall include any reasonable additional costs solely attributable to Owner's termination for convenience but shall not include any amounts for lost revenues, lost profits, or any other consequential damages.

### Owner can obtain documentation

The owner will want to be able to substantiate the contractor's entitlement to any position the contractor takes. Certain provisions relating to specific contractor requests—such as those relating to requisitions for payment, change orders, and claims—include a requirement for documentation. However, the owner should have the ability to obtain relevant documentation related to other issues as well. Exhibit 7.9 is an example of such a provision.

---

**Exhibit 7.9   Owner can obtain documentation**

13.1 *Books and Records* Contractor shall keep for at least six (6) years after final payment, books, records, accounts, and all documents created or received by Contractor in the performance of the Work or otherwise pursuant to this Contract. Owner, with reasonable notice, shall at any time during the progress of the Work and for such six (6) years after final payment have access to and may obtain copies from Contractor of such books, records, accounts, and documents. Owner shall reimburse Contractor for the actual cost of any copies of any documents exceeding in total ten (10) pages.

---

### Waiver

As discussed in Chapter Three, owners must be alert to avoid actions or failures to act that may have the effect of waiving contractual requirements. The construction contract should assist this effort by including a provision that seeks to minimize the acts or failures to act which actually constitute a waiver, as well as minimizing the effect of those that are deemed to be waivers.

Exhibit 7.10 illustrates such a waiver provision.

---

**Exhibit 7.10   Waiver**

14.1 *Waiver* No act or failure to act by Owner shall relieve Contractor of any obligation imposed on Contractor by the Contract Documents. Any relinquishment by
Owner of any right or waiver of any Contractor obligation shall be effective only
if it is in writing, signed by Owner and addressed to Contractor, and shall be
effective only for the specific act or omission addressed in such written waiver
and not for any other similar or different act(s) or omission(s).

---

## Reviewing designer-prepared technical specifications included in construction contracts

It is very important for the owner to review the technical specifications included in any
construction contract prior to issuing the contract for bid to make sure that they create clear
obligations for the contractor. The language can be very informal when it comes from the
designer. The objective of every provision of the specifications should be to create a clear
directive to the contractor. It should clearly describe the work to be performed. In certain
cases the designer may elect to create a standard to be met and leave it in the contractor's
discretion how the standard will be met (this is known as a performance standard). In such
cases the standard must be clear. Clear specifications minimize disputes.

People who are not architects or engineers should not get involved with the appropriate-
ness of technical requirements. However, non-designers can and should review the specifications to ensure that they constitute clear requirements. They should not be written as
hopes, expectations, anticipations, or similar language, none of which is binding on the contractor. Exhibit 7.11 illustrates the type of language to be avoided and Exhibit 7.12 shows the
appropriate language.

---

**Exhibit 7.11   Informal specification provision**

It is anticipated that windows in all offices in the best locations shall be operable.

---

There are three problems with this specification. The first is its use of the words "it is
anticipated." This is not a requirement. If the contractor does not provide any operable
windows, it can argue that while operable windows may have been anticipated they were not
required because anticipation represents an expectation but not a requirement. The second
problem with the specification in Exhibit 7.11 is it does not identify the offices getting the
operable windows with any specificity. This makes the identification of these offices a matter
of the contractor's discretion. The third problem is that the specification does not describe
which type(s) of operable windows are to be installed. The specification in Exhibit 7.12
addresses each of these problems.

This provision establishes a requirement, specifically describes the offices that will receive
the operable windows, and identifies which windows are acceptable.

---

**Exhibit 7.12   Correctly written specification provision**

Contractor shall install operable windows in all corner offices and all other offices with at least four hundred and fifty (450) gross usable square feet as depicted on drawings A____, A____, and A____. Such windows shall be Anderson, Marvin, or approved equal.

---

## Provisions unique to construction manager at risk contracts

This section discusses provisions that are unique to construction management at risk (CMR) contracts. These provisions, which are typical for CMR contracts in both the public and the private sectors, include the following:

* preconstruction services;
* guaranteed maximum price (GMP);
* direct cost of the work;
* general conditions costs;
* contractor's fee;
* contractor's contingency;
* GMP savings;
* GMP amendment; and
* work prior to GMP amendment.

Each of these issues will be discussed below.

While some CMR contracts refer to the entity performing the work as the "Construction Manager," others refer to the entity as the "Contractor." The discussion below will use the term "Contractor." Using the term "Contractor" in the contract puts an appropriate emphasis on the primary duty of the contracting party as building the project. It eliminates any emphasis on the bells and whistles that the CMR approach may add to the construction process. Using the term "Contractor" is also a recognition that many construction firms that label themselves construction managers, because that is a term with a lot of favorable appeal to owners, actually operate as general contractors from a business perspective. The third reason for using the term "Contractor" is that it eliminates any confusion about whether we are talking about a construction manager at risk (i.e., a contractor) or about a construction manager as agent (i.e., an owner's representative).

### *Preconstruction services*

Under a CMR contract, the owner generally procures the contractor soon after the designer. This approach provides the benefit of contractor involvement starting at an early stage of the project design. The contractor is often selected during the schematic design phase.

There are several benefits to involving the contractor early in the design process. The contractor can provide input on constructability issues. As design progresses, the contractor can provide a more comprehensive review of the completeness of the design documents. It can perform periodic estimates to determine if the design can be built within the project budget. If value engineering is necessary, the contractor can play an active role in developing effective proposals.

As design progresses, the contractor can also provide input into, or be primarily responsible for, how bid packages will be assembled. This responsibility has two aspects. The first is a determination of which portions of the work will be placed in each bid package. In many public jurisdictions, there are statutory prohibitions or limits on the extent to which the contractor, or the designer, can assemble bid packages. However, in the private sector, this opportunity is unlimited and an important part of the contractor's construction planning. The second aspect of assembling bid packages is the actual preparation of the bid package. Either way, the owner should pay close attention to the preparation of the bid packages to make sure that the design documents and the instructions to bidders in each bid package are complete, unambiguous, and don't contain inconsistencies.

One of the advantages of involving the contractor in the preconstruction phase is its relationships with subcontractors. The contractor can discuss design issues, constructability, and pricing with subcontractors that actually perform the type of work in question. Typically, the contractor will hold these conversations with one or more subcontractor with which it has done business in the past and in which it has confidence.

The owner can procure these preconstruction services in one of two ways. The first, and most common, is to enter into a single contract for the project which includes both preconstruction and construction services. An alternative procurement strategy is to contract only for preconstruction services, usually with an option, exercisable at the owner's discretion, for construction phase services. This approach allows the owner to limit its early financial commitment and to determine how comfortable it is working with the selected contractor.

### Guaranteed maximum price

The CMR type of contract is essentially a cost plus type of contract with the contractor's compensation capped at a not to exceed value, the guaranteed maximum price (the "GMP"). It is important to understand that the GMP is not a lump sum. Under a GMP contract, the contractor will be paid a total amount in accordance with contractual provisions covering compensation, and that amount may equal the GMP or may actually be less than the GMP. The difference, if any, is called the GMP savings and is discussed below. The GMP can also be exceeded if the owner agrees to change orders that increase the GMP.

The GMP is based primarily on three elements. These include the design documents, either the 100 percent construction documents or the most current edition approved by the owner; the schedule as approved by the owner; and the qualifications and assumptions. Each of these is included in the CMR contract as an exhibit. The design document exhibit is typically a list of all the included drawings and specifications. The schedule may be included in its entirety or, on a large project, incorporated by reference. The qualifications and assumptions are included in their entirety.

The design documents are prepared by the designer and approved by the owner. Only design documents approved by the owner are included in the contract. The schedule is developed by the contractor and approved by the owner. Only as approved will the schedule be included in the contract. The qualifications and assumptions are initially prepared by the contractor and then negotiated with the owner. Only the version that is mutually agreed upon is included in the contract.

Qualifications and assumptions, often called "Q&As," are theoretically intended to clarify ambiguities in the scope and/or the business deal. However, it is important for the owner to pay close attention to the Q&As because in most cases they represent modifications to the scope as expressed in the design documents or to the business deal as expressed in the contract. Exhibit 7.13 shows an example of an item that might be in the Q&As.

---

**Exhibit 7.13   Sample qualification and assumption**

The requirement in Section ___ that countertops shall be first quality granite shall apply only to domestic granite. If the Owner requires use of an imported granite, the extra cost of such imported granite over the cost of the equivalent domestic granite shall be reimbursed to the contractor as a change order.

---

As can be seen from the example, a clarification inevitably modifies the scope and/or the business deal. For that reason, owners need to pay very close attention to the draft Q&As as prepared by the contractor.

For those not familiar with this project delivery mechanism, it is important to understand that a GMP is no more guaranteed than a lump sum price. CMR contracts have change order provisions that are identical to the change order provisions of a lump sum contract. That is because, as described above, the GMP is based on scope, schedule, and Q&As and these can change.

It is important when drafting a CMR contract to clearly define the four components that make up the GMP: direct costs, general conditions costs, fee, and the contractor's contingency.

### *Direct cost of the work*

Direct costs are the costs incurred by the contractor in performing the work. These are mainly amounts paid to subcontractors because under the CMR format, the contractor usually does not self perform any of the work. Nonetheless it is important to specify in detail what constitutes acceptable direct costs for two reasons. First, this provision will flow down to the subcontractors through their subcontracts and thereby govern what costs they can claim. Second, in the rare instances when the contractor self performs work, the provision will govern the costs that the contractor can claim as direct costs. Exhibit 7.14 illustrates a provision that enumerates direct costs.

---

**Exhibit 7.14   Direct cost of the work**

The term Direct Cost of the Work shall mean those costs necessarily incurred and actually paid by Contractor in connection with the performance of the Work in accordance with the Contract Documents for the following items:

1   wages and benefits of trade workers directly employed by Contractor performing the construction of the Work at the Site;
2   reasonable rental or ownership costs for construction equipment other than small tools used in the performance of the Work, and associated costs such as fuel, installation, dismantling, and minor repairs, provided that Owner may require that it approve in advance the billing rates for all or specific pieces of equipment;

---

3  sales and use taxes, permit fees, royalties, and deposits lost for causes other than Contractor's negligence;

4  costs of repairing damaged or Defective Work provided that such damaged or Defective Work did not result from the fault or negligence of Contractor or any Subcontractor and only to the extent such costs are not recoverable from others or are not reimbursed by insurance or otherwise;

5  actual amounts paid (inclusive of discounts) for materials, supplies, and equipment installed or to be installed in the Project, including transportation and all other costs required for the proper furnishing of such items;

6  the amounts paid by Contractor to any Subcontractor in accordance with the terms and conditions of this Contract and any approved Subcontract;

7  costs incurred in taking action reasonably required by an emergency to protect against bodily injury or property damage;

8  costs of premiums for insurance polices required by this Contract;

9  premiums on all bonds required by the Contract Documents, less all refunds of such premiums;

10  legal costs, including attorneys' fees, other than those arising from disputes between the Owner and Contractor, reasonably incurred by the Contractor in the performance of the Work with the Owner's prior written approval; and

11  other costs incurred in the performance of the Work if and to the extent expressly approved in writing by Owner.

In no event shall any item of the Direct Cost of the Work include or be based on rates that are higher than the standard paid at the place of the Project except with prior written approval of Owner.

### General conditions costs

General conditions costs are essentially overhead costs. These costs include items that support the performance of the work but do not involve actual performance of the work. Typical examples include the contractor's project management team, the trailer(s) in which they are located, and the supplies and equipment for the trailer. If the contractor has a crew of laborers which does such things as assemble scaffolding, clean up the project site, operate a wheel wash, and/or other tasks that support construction activities, this crew would be part of general conditions. Exhibit 7.15 is an example of a provision that describes general conditions costs.

An important drafting issue is how the general conditions costs are to be priced. One alternative is to establish a set amount for general conditions costs. The other alternative is to establish a rate for general conditions costs. This rate, expressed as a percentage, is applied to the direct cost of the work to determine its value. For example, in the case of a project with $20 million of direct costs, where the contractor's general conditions cost is determined by using a 6 percent rate, the total general conditions costs, assuming no change orders, would be $1.2 million.

The advantage of the set amount for the owner is that it provides a strong incentive to the contractor to live within the amount established. If the value of the general conditions costs is established by applying a percentage to the direct cost of the work, there is an incentive

---

**Exhibit 7.15   General conditions costs**

General Conditions Costs shall include all costs necessarily incurred and actually paid by Contractor for the following items:

1   wages and benefits for workers directly employed by Contractor in the performance of the Work, but not including the wages and benefits paid to workers employed by Contractor and performing trade work as authorized by Clause 1 of Paragraph ___ [the paragraph describing Direct Cost of the Work];
2   salaries and benefits of Contractor's management, supervisory, and administrative personnel assigned to the Project;
3   all expenses associated with travel by the Contractor's management or other personnel in connection with the Project, including relocation of any personnel, provided that relocation costs shall only be paid pursuant to advance written approval by Owner;
4   costs for furnishing, including transportation, installation, maintenance, repair, replacement, dismantling, and removal, of materials, supplies, temporary facilities, machinery, vehicles, equipment, and hand tools, which are provided by Contractor at the Site for use to perform General Conditions work [i.e., work required by the General Conditions and therefore not trade work];
5   costs for performing clean-up and legal removal of debris from the Site;
6   costs of providing the Site office (including field offices for the Owner's and Designer's representatives) and associated services and facilities, including facsimile and local and long-distance telephone service, computers and associated equipment costs, printing and reproduction of documents, messenger services, postage and parcel delivery, and petty cash expenses of the Site office;
7   connection and usage fees for temporary lighting, electricity, heating, water, and any other utilities;
8   costs associated with development and administration of any drug testing, rodent control, and/or other programs required by the Contract Documents that will necessarily involve third party vendors; and
9   any other costs incurred and actually paid by Contractor that are related to the management of the Work with advance written approval by Owner.

---

for the contractor to let direct costs increase, or, at a minimum, less of an incentive to control general conditions costs.

How the amount due for general conditions costs for a given requisition would be calculated is explained below. In each case the effect of change orders will not be considered. This issue will be discussed in Chapter Nine where mark-ups on change orders are specifically addressed.

If the costs are determined by a percentage, then the percentage is applied to the total value of the direct cost of the work billed to date; then from that value is deducted the value of total payments to date for general conditions costs, leaving the amount due for this requisition. An alternative is to calculate the amount due for direct cost of the work for the requisition and apply the percentage to that amount. Since both methods should produce the same value, either method should serve as a useful check on the other.

To continue the example of the $20 million project, if $14 million of direct cost of the work has been earned, the total general conditions cost payable is $840,000 ($14,000,000 × 0.06). If the total amount of general conditions costs paid to date is $795,000 (based on payment of $13.25 million of direct cost of the work), the amount due for the current requisition for general conditions costs is $45,000 ($840,000 − $795,000). Similarly, if $13.25 million has been paid for direct cost of the work, that means $750,000 is due for direct cost of the work ($14,000,000 − $13,250,000) for the current requisition; and, applying the 6 percent rate, the amount due for general conditions costs is $45,000.

If general conditions costs are a set amount, the calculation is based on percentage completion of the work. If on the $20 million project, the set amount for general conditions costs is $1 million, and if the work is 40 percent complete, and 35 percent has previously been paid, the total owed for general conditions costs would be $400,000 ($1,000,000 × 0.40), the amount previously paid would be $350,000 ($1,000,000 × 0.35), and the amount owed for the current requisition would be $50,000 ($400,000 − $350,000).

## Contractor's fee

The contractor's fee is the contractor's profit. The actual cost of performing the construction work is paid for as the direct cost of the work. The project overhead is paid for as the general conditions costs. This leaves the contractor's fee to be the source of the contractor's profit for the project.

The contractor's fee has two other functions. A portion of the fee is used to fund corporate overhead. That is true for all of the contractor's projects, regardless of delivery method; each project must contribute to corporate overhead. The other function of the contractor's fee is to cover extra costs that cannot be reimbursed through change orders. This use of fee is strenuously resisted by most contractors because it reduces the amount available for profit and contribution to corporate overhead.

Like reimbursement of general conditions costs, the contractor's fee can be paid for as a set amount or as a percentage. As in the case of general conditions costs, defining the contractor's fee as a percentage reduces the contractor's incentive to control the underlying costs (i.e., the direct cost of the work and general conditions costs).

If the contractor's fee is paid as a percentage, the percentage is applied to the sum of the direct cost of the work plus the general conditions costs. So, to continue the example of the $20 million project from the general conditions discussion, if the fee is set at 3 percent, the total fee owed to the contractor, assuming no change orders, will be $636,000 calculated as follows:

| Direct costs | = | $20,000,000 | | |
|---|---|---|---|---|
| General conditions costs @ 6% | = | $20,000,000 × 0.06 | = | $1,200,000 |
| Direct costs plus general conditions costs | = | $21,200,000 | | |
| Fee @ 3% of direct costs plus general conditions costs | = | $21,200,000 × 0.03 | = | $636,000 |

The calculation of the amount of fee owed on a particular requisition is calculated in the same way as the general conditions costs.

If the contractor's fee is a set amount, the calculation for a specific invoice will be similar to the calculation for general conditions costs where those costs are a set amount. The amount of the fee payable as part of any requisition will equal the total fee times the percent of the work completed minus the amount of the fee previously paid.

### Contractor's contingency

The contractor's contingency is included in the GMP to cover unforeseen events. Every lump sum bid also has a contingency for unforeseen events. What makes the CMR contract's approach different is that it makes the contingency transparent. It does this in four ways.

First, the contract specifies the types of events or circumstances as a result of which the contractor is authorized to use the contingency. Second, the contractor's contingency is a specific line item in the GMP budget as a "below the line" item (i.e., not in the GMP). Third, when the contractor uses contingency funds, each use is required to be included in the next requisition. Fourth, the contractor is required to track and report on the status of the contingency.

These aspects of transparency mean that the owner is actively involved in the management of the contingency. That involvement can become significant if the contract provides the owner the ability to approve specific uses of the contingency. Typically, under a CMR contract, the contractor starts out in control of the contingency, but the owner is given the authority to require its approval for future uses of the contingency either at its sole discretion or if certain contractually specified events or conditions occur.

The other important issue concerning the contractor's contingency is what happens if at final completion there is an unspent balance of the contingency. As discussed in more detail in the next section, some or all of the unused contingency is returned to the owner. This leads many owners to consider the contractor's contingency their money—or at least potentially their money—and therefore to resist contractor uses of the contingency.

The net effect of these various provisions is to get the owner actively engaged in the use of the contingency. Furthermore, there are incentives for the owner and the contractor to fight over the proposed uses of the contingency. The contractor wishes to use contingency to cover extra costs it cannot fund through a change order in order to protect its fee (i.e., not have to eat into its fee to cover unforeseen costs for which it can't get paid by the owner). The owner is interested in minimizing the use of the contingency in order to maximize the amount available to the owner at the end of the project.

The contractor's contingency in a lump sum bid is hidden. There is only one value—the lump sum—disclosed to the owner and so the contingency, like the various subcontract values, are not disclosed. (Under various public procurement procedures, certain subcontract values may be itemized but not contingency.) The contractor uses its contingency when and as it sees fit. It is a matter solely in the contractor's discretion and is not reported to the owner. Therefore, in a lump sum situation, the owner and the contractor have no reason to discuss use of the contingency.

Under a CMR contract, when the contractor uses contingency funds, they are then added to the GMP by change order (because the contract says the GMP can only be increased by change order). Usually the amount of the contingency used is marked up by the same percentages for general conditions and fee as other change orders.

A particularly challenging issue for those not used to CMR contracts is use of the contingency to pay for the correction of defective work. There is a temptation to resist using the contingency for this purpose. However, the contractor needs a pool of funds to pay for the correction of defective work. Furthermore, in a lump sum contract, the contractor has, as discussed above, included a contingency which is invisible to the owner. One purpose of that contingency is to fund correction of defective work. Most CMR contracts limit the use of contingency funds to correct defective work to situations not involving contractor negligence.

### *GMP savings*

GMP savings is the term used to describe the difference between the GMP and the total amount owed and paid to the contractor pursuant to the contract, assuming that the number is a positive value. The GMP is a maximum price. Under a CMR contract, the owner pays costs up to the amount of the GMP. These costs, as discussed above, include only the direct cost of the work, the general conditions costs, and the contractor's fee. If the total amount of those costs payable to the contractor is less than the GMP the difference represents the GMP savings. For example, if the GMP is $5 million; the total direct cost of the work is $4,500,000; the total general conditions cost is $270,000; and the total contractor's fee is $143,000; there are GMP savings of $86,900 as shown in Exhibit 7.16

---

**Exhibit 7.16    Calculation of GMP savings**

| | |
|---|---:|
| GMP | $5,000,000 |
| Direct cost of the work | $4,500,000 |
| General conditions costs (@ 6%) | $270,000 |
| Subtotal | $4,770,000 |
| Contractor's fee (@ 3%) | $143,100 |
| Total costs | $4,913,100 |
| GMP savings (GMP − total costs) | $86,900 |

---

The key drafting issue is how the savings are split. The objective is to balance two competing incentives. On one hand, the larger the contractor's share of the savings, the stronger the contractor's incentive to save costs. On the other hand, the larger the share of the savings payable to the contractor, the more likely it is the owner will lose interest in sharing the savings at all.

It is not unusual for the owner to retain all the savings. This is particularly true for public owners. When the contract provides for shared savings, the contractor's share is rarely greater than 50 percent and almost never less than 25 percent. Exhibit 7.17 contains a sample savings provision based on the contractor getting 25 percent of the savings.

---

**Exhibit 7.17    GMP savings provision**

The GMP Savings, if any, shall be the amount by which the GMP exceeds the sum of the total of a) Direct Cost of the Work, b) General Conditions Costs, and c) the Contractor's Fee payable to Contractor. The GMP Savings shall be shared between Owner and Contractor as follows: seventy-five percent (75%) shall be retained by Owner and twenty-five percent (25%) shall be paid to Contractor as part of the Final Payment as authorized by Paragraph ___.

---

### GMP amendment

The contractor under a CMR contract is generally procured soon after procuring a designer. The important point from a contracting standpoint is that the design documents are not close to completion when the contract is negotiated, and, therefore, the contractor cannot realistically price the cost of building the project. For that reason, contractors are typically required to price preconstruction services, general conditions costs, and the contractor's fee as the cost part of their proposal. To the extent the procurement is competitive, and price is a factor in the evaluation of proposals, these are the values that are evaluated.

When the design documents become 100 percent complete (or some lower percentage established by the contract), the owner and the contractor need to agree on the GMP. The document that establishes the GMP at that time is commonly called the GMP Amendment.

This document usually does more than establish the value of the GMP. It typically includes a number of exhibits on which the GMP is based. These include at a minimum:

* a list of the final design documents;
* a construction schedule approved by the owner; and
* a list of the agreed upon qualifications and assumptions.

Sometimes after the project has gotten underway, the parties decide that one or more provisions of the contract and/or the general conditions should be modified. This amendment is sometimes used to effect those changes.

Often there will be additional exhibits that the owner and the contractor agree are necessary or helpful to define the project and/or the contractor's scope. When using additional exhibits, it is important to ensure that the exhibits contain clear requirements or specific information in order to avoid creating ambiguities or inconsistencies with other portions of the contract documents.

Some owners prefer not to enter into the CMR contract till the GMP is agreed upon. In such cases, the GMP Amendment will not be necessary.

### Work prior to GMP amendment

If the CMR contract contemplates the owner hiring the contractor during the early stages of design, then another drafting issue unique to CMR contracts may arise.

One potential advantage of the CMR approach is the ability to have the contractor start construction work before all the design documents are complete. The principal benefit of this approach is compression of the project construction schedule. The principal drawback is that the owner starts construction of the project without knowing with any real confidence what the GMP will ultimately be.

For those owners that wish to authorize early (i.e., pre-GMP) work, or to preserve the option to do so, they will want a provision specifically authorizing early work in the CMR contract. That provision should require owner authorization, and establish a process for selecting trade contractors to perform the work. A sample provision is shown in Exhibit 7.18.

---

**Exhibit 7.18   Early work**

Owner may authorize Contractor to perform one or more portions of the Work prior to establishment of the GMP ("Early Work"). The Early Work shall be procured based on one or more bid packages which shall be approved by Owner in writing prior to issuance. Early Work bid packages shall only be provided to subcontractors approved in advance by Owner. The Early Work shall be performed in accordance with the terms and conditions of the Contract.

---

## Conclusion

Drafting of the construction contract, whether design-bid-build or CMR, is an important part of the owner's effort to control scope, cost, and schedule. To do that effectively, the contract must appropriately balance risks, provide clear direction to the contractor, and protect the owner's interests. The provisions discussed in this chapter seek to accomplish those objectives. However, there are certainly many alternative provisions that will accomplish the same objectives.

# 8 Administering the construction contract

## Introduction

As discussed in Chapter Three, contract administration should focus on ensuring that the owner receives the deliverables promised in the contract. That approach applies to the construction contract. Some of the most important deliverables that contractors are usually required to submit include the following:

- schedule of submittals;
- schedules of values and requisitions for payment;
- baseline schedule and progress updates;
- quality control plan/quality control manager;
- safety plan/safety manager; and
- requests for substitution.

All of these deliverables are important to enabling the owner to control scope, cost, and schedule. The importance and appropriate use of these deliverables should not imply that the process of developing, submitting, reviewing, approving, and using these documents must be adversarial. Their development, processing, and use can be, and often are, the subjects of a collegial approach involving the owner, the designer, and the contractor and their respective consultants.

Each of the listed deliverables is discussed below. The purpose of the deliverable and why it is important to the owner is addressed. The chapter concludes with a recommended approach for administering the construction contract at the start of the project to maximize the likelihood of a successful project.

## Schedule of submittals

The schedule of submittals is a list of all the deliverables required by the construction contract to be submitted to the owner. This document allows the owner and the contractor to agree on what deliverables are required by the contract and when they are required to be submitted. The contract should require this schedule to be complete. It should include all the deliverables required by the contract. These include technical submittals such as shop drawings, product samples, and the construction schedule and updates; management submittals such as quality control and safety plans; and business submittals such as requisitions for payment and information on proposed subcontractors and proposed subcontracts.

The schedule should show when each submittal will be made. Recurring submissions, such as schedule updates and requisitions for payment, can have their submittal dates shown as certain days of the month (e.g., the fifth business day of the month) or as specific dates (e.g., March 5, 2016).

## Schedule of values and requisitions for payment

The schedule of values breaks the construction work into specific activities and assigns a value to each activity. The total of the activity values must equal the value of the construction contract. Then, as the construction progresses, the schedule calls for additional information on each activity, including the percentage of the work completed, the total value of the completed work, the amount paid to date, the amount due for the current pay period, and the value of the work yet to be performed. The schedule of values is usually required to be updated and submitted monthly with the requisition for payment.

The schedule is updated as change orders are approved. This involves increasing or decreasing the values of activities on the schedule. It also includes adding new activities and their values as new scope is added, and/or deleting activities currently on the schedule and their values as scope is deleted.

The schedule of values is a very important document for both the contractor and the owner. That is because the schedule of values determines the amount of each requisition for payment. The basic calculation is described in Exhibit 8.1.

---

### Exhibit 8.1   Calculating monthly requisition for payment

The monthly requisition for payment is calculated as follows.

1   For each activity the contractor establishes the total value of the work performed to date by multiplying the total value of the activity by the percentage of the work completed to date.
2   From the value of the work performed to date is deducted the amount paid to the contractor to date for that activity, leaving a balance to be paid for that month.
3   The balances to be paid for all activities for that month are added up to reach a total amount due for that month.

---

The percent complete for each activity is what drives the monthly calculation because it is percent complete that determines the total amount owed to the contractor to date for that activity. The second and third steps are straightforward mathematical calculations. For this reason, the owner needs to closely monitor the percent complete claimed by the contractor for each activity on the schedule of values. Determining a percent complete is usually a matter of judgment. An important control mechanism is the use of construction inspection performed by or on behalf of the owner so that the owner can have an independent assessment of the percent complete of all items billed for in a particular requisition for payment.

The schedule of values is a very important document from the stand point of administering the construction contract. That is because the schedule of values is the owner's primary mechanism for preventing the contractor from "getting ahead" of the work. "Getting ahead" in this context means that the amount the contractor has been paid on a given date significantly exceeds the actual value of all of the work performed to that date. If the contractor gets ahead in this sense, it provides negative incentives to the contractor. For example, the contractor may use the excess amount paid to fund other jobs. This may result in the contractor having insufficient funds to perform the owner's work correctly, or, in extreme cases, to perform the work at all. Another, more common, version of this problem is that the contractor fails to perform the punch list work because there is no longer enough money in the contract to fund the work.

The contractor will try to get ahead in one or both of two ways: front end loading and inflating percentages of completion. Front end loading refers to assigning dollar values to the first several activities on the schedule of values that significantly exceed the actual value of performing the work. The way to mitigate this problem is to carefully evaluate all values proposed by the contractor in its initial proposed schedule of values, with particular attention to early activities.

The second way a contractor may try to get ahead is to inflate the percent of completion of one or more activities. The principal mechanism for dealing with this approach is regular and detailed inspection of the work by the owner. If the owner has independent knowledge of the progress of construction, it can effectively resist the contractor's attempt to overstate the value of work performed to date.

Requisitions for payment are important because they are the mechanism by which the contractor gets paid. If the schedule of values as submitted each month with the requisition is carefully reviewed as described here, the owner's review of the requisition will be primarily focused on the adequacy of the supporting documentation and on mathematical accuracy.

## Baseline schedule and progress updates

The typical construction contract will require the contractor to submit and update a construction schedule. Chapter Nine includes a brief discussion of construction schedules in the context of discussing defending against change order proposals claiming delay damages. This section is concerned with the administration of the contractual requirements related to scheduling.

The construction schedule is of great importance to the owner. It represents the contractor's plan for completing the project by the contractually specified completion date. Monthly updates are important because they indicate whether the plan is working (i.e., whether the project will finish on time). If there are reasons the project may not finish on time, they will become apparent from the monthly updates.

The contract will specify the type of schedule: CPM or bar chart. It may specify characteristics of the schedule. Examples of these requirements include the minimum or maximum number of activities and the maximum duration for a single activity. For bar chart schedules, the contract may specify if the unit of time across the top of the schedule should be weeks or months. For CPM schedules the contract may require a network diagram, indicate how the critical path is to be illustrated or described, preclude so-called dummy activities, and require cost loading.

Regardless of the type of schedule required, the contract will indicate when the contractor must submit the initial schedule to the owner. Usually the submission deadline is stated in number of days following award of the contract. The initial schedule is often called the baseline schedule because it represents the "baseline" against which the progress of the project is measured.

Administration of the scheduling requirement begins with getting the contractor to submit its initial schedule within the contractual time frame. This is not always easily done. If necessary, the owner should be prepared to make repetitive, documented requests for the schedule. This effort should be supported by a contractual provision that makes an approved baseline schedule a precondition to the contractor receiving its first progress payment. The scheduling provision in the construction contract in Appendix C in the Appendix section at the end of the book includes this requirement.

The next step is determining if the proposed schedule complies with the contractual requirements related to form. Examples of contractual requirements related to form include whether the schedule is a bar chart or a CPM schedule, a limit on the number of activities shown, and whether the schedule must be cost loaded. If it does not comply with the requirements, it should be returned to the contractor with a transmittal detailing the deficiencies and giving the contractor a deadline for submitting the corrected schedule.

When the schedule meets all contractual requirements, the next step is to determine if the schedule represents a realistic plan for completing the project by the contractually specified completion date. This is the standard established in the construction contract in Appendix C for approving the contractor's proposed schedule. What this standard means is that the owner's sole inquiry is whether the schedule describes a realistic way for the contractor to complete the project. If the answer is yes, the schedule should be approved, even if the owner (or its consultants) believes there is a better way to build the project.

The owner should only reject the contractor's proposed schedule if it believes the schedule is unrealistic. An unrealistic schedule might contain such things as activities with durations that are too short, activities shown as concurrent which can't be performed simultaneously, and/or missing activities necessary to build the project. If it rejects the contractor's proposed schedule, the owner should provide a written explanation of the reason(s) for the rejection. When the owner is prepared to approve the schedule, it should do so in writing clearly identifying the version it has approved.

Exhibit 8.2 summarizes the steps for approving the baseline schedule.

---

**Exhibit 8.2    Steps for approving the contractor's baseline schedule**

The following are the steps involved in approving the contractor's proposed baseline schedule.

1    Obtain the schedule from the contractor.
2    Determine if the proposed schedule meets all contractual requirements. If it does not, return with written comments for further development by the contractor.
3    Determine if the schedule is a reasonable plan to complete the project on time. If not, return with written comments for further development by the contractor.
4    When appropriate, approve the plan in writing.

---

Most construction contracts require monthly updates of the schedule once construction begins. There are three administrative tasks associated with these updates. The first is obtaining the monthly update from the contractor. The second is making sure the update complies with any contractual requirements for monthly updates. The third is determining if the update is an accurate reflection of the project's status. This determination should be made in close coordination with the owner's construction inspector.

Written communications concerning the contractor's baseline schedule and updates should always be as detailed as possible. If the submission does not comply with certain contractual requirements, those requirements should be cited or quoted. If the submission is judged not to be a realistic plan (baseline schedule) or an accurate description of the status of the project (monthly update), the reason(s) for the conclusion should be described in detail. This approach makes it easier for the contractor to improve its submissions. It also creates a clear record of failure to comply with the contract if that later becomes relevant.

## Quality control plan/quality control manager

It is useful to define two terms related to quality that are often used interchangeably. The contractor is responsible for the quality of its work. That is quality control, often called "QC." It is the owner's responsibility for making sure that the contractor implements quality control. That is quality assurance, often referred to as "QA." This section discusses administering the provisions of the construction contract that relate to the contractor's responsibility for quality control.

Quality control is important to the owner because it is the function that ensures that the physical work complies with the contract requirements. The owner will want to be sure the contractor is carrying out its quality control responsibilities. The level of quality for which the contractor is responsible is that specified in the contract documents (not the level the owner, designer, or a consultant decides is applicable on an ad hoc basis). The construction contract should require the contractor to submit a quality control plan. The contractor's quality control plan will describe how the contractor intends to make sure that its work meets the quality requirements of the construction contract. It would typically address such topics as inspection, nonconformance, the nonconformance log, correction of deficiencies, and reporting to the owner.

Quality control starts with inspection. The plan will describe the personnel assigned to inspect the work and the manner and frequency of their inspections. The plan will next address what happens when defective work is observed. Typically this involves filling out a nonconformance report, often referred to as an "NCR." On larger, more complicated projects, the owner may want to review the contractor's proposed NCR format.

The NCRs should be numbered for tracking purposes. The form should allow for the deficiency to be described in sufficient detail, including the circumstances that caused the deficiency. The NCR should show the entity and which employee of that entity is responsible for correcting the deficiency. The NCR should also show the date by which the deficiency is to be corrected. It needs a place for the appropriate contractor manager, usually the quality control manager, to initial approval of the corrected deficiency. The owner may require additional approvals. Usually these would be by the designer and/or the owner's representative. The elements of a nonconformance report are shown in Exhibit 8.3.

---

**Exhibit 8.3   Elements of a nonconformance report**

A contractor's nonconformance report acceptable to the owner should have the following elements.

1   A number for tracking purposes;
2   enough space to describe the deficiency in sufficient detail;
3   the entity and employee of that entity responsible for correcting the deficiency;
4   the date by which the deficiency will be corrected;
5   a signature line for the appropriate contractor manager; and approval lines for the designer and/or the owner's representative.

---

The quality control plan will describe an NCR log to be maintained by the contractor. Typically, the log will list each NCR, when it was prepared, a brief description of the deficiency, its current status, and the date by which it will be corrected. Copies of the log are usually submitted to the owner monthly so that the owner can actively monitor the contractor's progress in correcting defective work.

The quality control plan should require the contractor to provide a quality control manager to be assigned to the project. There are two important issues. The first is the quality control manager's qualifications. That person should have demonstrated previous quality control experience on projects of similar size, scope, and complexity.

The second issue is the level of effort the owner expects. Many contractors use the term "dedicated" to mean the same thing as "assigned." In this circumstance, a quality control manager may be assigned to several projects simultaneously. Therefore, if the owner concludes that a particular project is best served by a full-time quality control manager, the construction contract should specify such a requirement.

### Safety plan/safety manager

Safety should be a top owner priority for at least two reasons. First, it is the right thing to do. The physical welfare of those working on the project should be a first priority concern of both the owner and the contractor. Second, the safer the job, the more likely it is to finish on time and the lower associated insurance costs will be. These savings may not be directly realized on a small project with a short duration. On larger, multi-year projects, these savings can be significant and can be quantified and realized by the owner.

Before addressing how safety will be achieved on its project, the owner should focus on where primary responsibility for safety should be assigned. It is the contractor that is building the project, and the people actually performing the work are the employees of the contractor or its subcontractors. Therefore, it is appropriate (and standard) that the contractor be contractually assigned responsibility for project safety under the construction contract. Exhibit 8.4 illustrates such a provision.

A major obligation of the contractor should be the submission of a safety plan for the owner's approval. That plan should detail how the contractor will provide a safe work

---

**Exhibit 8.4  Contract provision assigning responsibility for safety to the contractor**

3.18  *Safety* Contractor shall be responsible for all aspects of safety on the project. It shall be Contractor's responsibility to avoid, to the maximum extent feasible, bodily injuries and property damage resulting from performance of the Work. At least thirty (30) days prior to commencing work at the Site, Contractor shall submit to Owner for Owner's approval, a safety plan indicating how Contractor shall maximize safety on the Project, and Contractor shall not commence work at the Site until such safety plan is approved in writing by Owner. Owner's approval for the purposes of this Paragraph 3.18 shall be limited to determining that Contractor's safety plan represents a reasonable approach to ensuring safety on the project. Contractor shall designate a key member of its project team as Safety Manager. The Safety Manager shall be responsible for implementing Contractor's safety plan.

environment; how it will anticipate and avoid problems, accidents, and injuries; and how it will effectively deal with those that occur. Larger, more sophisticated contractors typically have corporate safety manuals. The owner needs to make sure the contractor has a project-specific plan.

A comprehensive safety plan should, at a minimum, address the following issues:

- *Safety manager* The safety plan should identify the safety manager and discuss the level of effort that the person will devote to this project. It should describe the safety manager's specific duties.
- *OSHA compliance* The plan should list the applicable U.S. Occupational Safety and Health Administration (OSHA) requirements and explain in detail how the contractor will ensure that each requirement is satisfied.
- *Trade by trade measures* The plan should specify in detail the necessary safety measures for each trade. It should also clarify the responsibilities of each subcontractor performing work on the project for implementing those measures.
- *Project security* The plan should address project security in terms of protecting the project as well as protecting persons not involved in the project. Protecting the project usually involves measures to make sure equipment and materials are not stolen or damaged. Protecting persons not involved relates to preventing the injury, or even death, of persons who don't work on the project but want to enter the site out of interest, curiosity, or intent to commit a crime. Both of these interests are addressed by such measures as fencing, security guards, and/or security dogs.
- *Drug testing* A key indicator of a safe workplace is the absence of alcohol and drugs. The safety plan should address whether there will be drug testing on the project, and, if not, what steps will be taken to ensure that workers on the project are not under the influence of drugs or alcohol. If there is to be drug testing, the plan should describe the testing protocol, including the procedure for obtaining samples of testable material (usually urine), the method of testing, and the appeal rights of persons who test positive.

An important part of the safety plan is the provision of a dedicated safety manager. As in the case of the quality control manager, if the owner believes that its project requires a full-time safety manager, that requirement should be included in the construction contract. For example, the next to last sentence in the sample provision in Exhibit 8.4 might read:

> Contractor shall provide a key member of its project team as a full-time safety manager and that person shall have no other responsibilities with respect to this or any other project.

The appropriate role for the owner with respect to safety is to monitor the contractor's compliance with its contractual obligations related to safety. In general, this means approving the contractor's safety plan and then monitoring the contractor's adherence to the provisions of the plan. The owner's ability to successfully monitor the contractor's compliance depends on the owner having the capacity to inspect the contractor's operations in the field. This is one more return on the investment in inspection capacity.

It is critically important that the owner not assume the contractor's obligation for safety. If the owner inadvertently or intentionally assumes that obligation, depending on the actual circumstances, the owner may well end up with the liability for any accidents, injuries, and/or damages to property that occur as a result of unsafe activities.

The owner should take two important steps to avoid assuming the contractor's obligations for safety. The first is to insert a provision in the construction contract which makes clear that the contractor has exclusive responsibility for safety. The provision in Exhibit 8.4 illustrates that approach. The second is to ensure that the person monitoring the contractor's compliance with the safety plan on behalf of the owner is well trained as to what constitute safety plan violations and as to exactly what to do when confronted with a safety violation. That person should understand that his/her responsibility is to immediately contact the contractor and alert it that its safety plan is not being followed. The initial notification should be a call to the contractor's superintendent or project manager and should be detailed as to the nature of the violation, the location, and which entity is involved. The call should be followed by a written confirmation (email is acceptable for this purpose) restating the timing, specifics, location, and entity involved.

If the contractor does not promptly cure the violation, a more formal written notification should be sent to the contractor. It should be addressed to the most senior executive of the contractor assigned to the project. If the contractor fails to avoid or promptly cure a major safety violation, or to promptly cure repeated violations of whatever seriousness, the owner should set up a meeting with the contractor's senior executive assigned to the project and the project manager. The agenda for the meeting should include:

• review of the problem in detail;
• review of contractor's contractual obligations for safety;
• discussion of how the contractor will improve its safety performance; and
• consequences of failure to improve.

If an authorized representative of the owner believes someone is in imminent danger of being injured or killed because of construction operations, he/she should stop the threatening work. Such an action should not be taken lightly, and only if there is not time to involve the contractor.

## Requests for substitution

Requests for substitution involve requests by the contractor to utilize different materials or to furnish and install different equipment than what is called for in the design documents included in the construction contract.

Contractors usually submit requests for substitution for one of two reasons. First, the contractor, or the subcontractor whose work is involved, believes the alternate material or equipment will actually improve the owner's project. The second reason contractors submit requests for substitution is because the alternative material or equipment can be purchased and installed by the contractor at a lower cost than it included for the item in the contract price. In the case of a lump sum contract, if the substitution is accepted, the contractor pockets the difference as pure profit.

The owner must effectively manage the process of evaluating requests for substitution so that requests which are based on the contractor's expertise and actually benefit the project are accepted, and those that would not benefit the project are rejected. There are three risks associated with requests for substitution. They are impacts on scope, impacts on cost, and impacts on schedule. Each request for substitution needs to be considered in terms of these three potential impacts.

The potential impact on scope is that the proposed alternative material or equipment will not perform in the same manner as the originally proposed item and that the difference will be significant enough to detract from the owner's original intent for the project. These differences can be particularly pronounced in the exterior and interior building finishes and in building systems.

The potential impacts on cost exist at three levels. The first potential impact is that the proposed material or equipment will cost more to purchase. The second is that it will cost more to install. Installation costs are typically labor costs so increased installation costs would usually mean it takes more trade worker hours to install the proposed substitute item. The third potential impact on cost is the additional cost of preparation to install the item. This may result from the proposed item requiring greater structural or other support than the original item. It may result from other dependent work having to be done in a more costly manner. Depending on the timing of the request for substitution, it could cause additional cost because it would require damage to, or even removal of, previously installed work, which would then have to be repaired or replaced.

The potential impacts on schedule can occur in three different ways. First, the proposed material or equipment may have a longer lead time from order to delivery than the original item. Second, the proposed item may take longer to install than the original item. Third, depending on the timing of the request, there may be additional time needed to repair or replace already installed work.

There are five steps for effective management of the request for substitution process.

The first step is developing a clear owner's program. It is the owner's program that establishes the goals for the project. For example, if the building is an office building, the key question is for what type of tenants the building is being built. Is the building intended to be an incubator for small hi-tech companies, a medical office building, or a class A building aimed at professional service firms? The answer to this and other questions establish the owner's goals for the building and, in turn, the appropriate design for that particular building. Requests for substitution should be considered in the context of those goals.

The second step is ensuring that the construction contract provides the owner with appropriate tools with which to manage the request for substitution process. The contract provisions addressing this process must give the owner the ability to avoid or control all three of the types of risk described above; that is, potential impacts on scope, cost, and/or schedule. Exhibit 8.5 is a sample of such a provision.

---

**Exhibit 8.5　Provision dealing with requests for substitution**

8.5　*Requests for Substitution* Contractor may submit a request for substitution ("Request") in connection with any specified material or equipment. The Request shall be submitted to Designer in writing at least thirty (30) days prior to the commencement of any construction work on the Project. The Request shall only be approved if the substitution would not directly or indirectly cause an increase in the Contract Price or an extension in the Contract Time and if Designer determines that the proposed material or equipment would perform in substantially the same manner as the originally specified material or equipment and that the substitution would benefit the Project. Owner, in its sole discretion, may instruct Designer to approve or reject the Request regardless of whether it meets the requirements of this Paragraph. Designer shall issue a written decision within twenty-one (21) days of receipt of the Request.

The third step in considering a request for substitution is to utilize all available expertise. If the owner will occupy the facility, the business units that will utilize the building, or floor, or laboratory are likely to have relevant expertise. For example, the owner's IT department may know a lot about the design and construction of data centers and therefore whether a particular request for substitution should be approved or rejected. If the owner is a developer, its leasing department and/or the brokers that lease or sell the developer's projects will have extensive knowledge of market expectations for that type of project and the impact of the proposed substitution on those expectations.

Another source of expertise is the project designer. Presumably the designer was retained because of its expertise in the particular type of project. The architect and/or the relevant engineer will have expertise that will be useful in considering a request for substitution.

The fourth step in managing requests for substitution is effective use of contractual tools. This becomes particularly important in cases where the contractor is inclined to be insistent about its request for substitution. In those cases the owner will want to know how to use its contractual tools to achieve the desired resolution. This can be rejection of the request, but, in other circumstances, it may be a significantly modified request which is acceptable to the owner.

Under the provision in Exhibit 8.5, the owner starts with full leverage because the provision reserves for the owner complete discretion to reject the request for substitution. If the owner is willing to consider some version of the proposed substitution, just not the one initially proposed, the owner can take the position that it will reject any version of the proposed substitution that is not what it wants.

The fifth step in managing requests for substitution is proactive monitoring of the item's installation. The monitoring of the actual installation is aimed at avoiding the contractor claiming as extra work costs which are actually indirect costs associated with the substitution. Those are costs for which the contractor would be responsible. Because the provision provides that the contractor is responsible for any additional direct or indirect costs attributable to the approved substitution, it is important to determine if there is work required that would not have been required absent the substitution.

An example will illustrate the point. If the owner approves a substitution for specified windows in a rehab project, any additional cost of the windows and window frames over the cost of the originally specified windows and frames would be direct costs. If the wall surrounding one or more of the window frames would have to be modified to accommodate the window frames, that would be an indirect cost. Under the provision in Exhibit 8.5, the contractor would be responsible for both of these costs. The purpose of monitoring the actual installation is to assist in determining whether the modifications to the wall, or any other work performed by the contractor and claimed as extra work, is associated with the substitution of the windows. For this purpose, work is associated with a substitution if it is caused by or required by the substitution. If it is associated, then it's an indirect cost for which the contractor is responsible. If it is not associated, then, depending on the circumstances, it may or may not be extra work for which the contractor is entitled to additional compensation.

## How to start the contractual relationship

As a starting point, to maximize the likelihood that the contractor will cooperate over the duration of the project, the owner should begin the construction phase of the project by rigorously enforcing contractual provisions. This establishes the owner's clear expectation that it will receive the physical construction and other deliverables at the cost and within the time for which it has contracted. It also sets a tone that the owner is serious about enforcing the contract. In most cases, this will engender a similarly businesslike response from the contractor (even if it complains).

# 9 Change orders and claims

## Introduction

This chapter addresses change orders and claims. Effective management of change orders and claims is vitally important to controlling scope, cost, and schedule.

Claims under many construction contracts, including the one in Appendix C, are required to be based on rejected change order proposals. In effect, they ask the owner to reconsider its earlier rejection based on changed circumstances, additional information, and/or a more extensive explanation of the basis of the original proposal. Under other construction contracts, all contractor proposals that are not requested by the owner are called claims. In either case, the merit determination, calculation of value (if relevant), and analytic techniques are the same as for the equivalent change order proposals. For that reason, the focus of this chapter is primarily on change order proposals and, except for a brief final section, claims are not discussed separately in this chapter.

Change order management should be viewed as a key component of the owner's cost control program. That is because the final value of a contract is rarely the same as the original contract price. Change order management is an important part of the owner's effort to minimize increases in the contract price.

Three principles should guide the owner's change order (and claims) management program.

- The contractor should be treated fairly.
- The owner should not agree to any change orders or claims to which the contractor is not entitled.
- Where there is entitlement, the owner should only agree to the amount of compensation (money and/or time) that is justified by the extent of the entitlement.

This chapter is aimed at explaining how the owner implements these principles. It first discusses certain particularly important provisions of the construction contract related to change orders. Then, it explains the contractor's burden of proof with respect to change order proposals, starting with requirements that all proposals must satisfy and then explaining the requirements for 12 of the most common types of change order proposals. The chapter next explains seven of the most common owner defenses to change order proposals. The following section explains how to calculate the appropriate value of a change order that has merit. The next section offers techniques for analyzing proposals. The final section briefly addresses claims and discusses the limited ways in which they may differ from change order proposals.

## Contract provisions

An owner's effective change order management program starts with favorable contract provisions. There are six particularly important issues.

- *Notice* It is important that the owner receive timely notice when the contractor believes an event or circumstance entitles it to additional money and/or time.
- *Required documentation* The contractor should be required to satisfactorily document its request for a change order.
- *Unilateral change orders* The owner should be contractually authorized to issue unilateral change orders.
- *Damages for delay* The contract should clearly establish the limits to allowable damages for delay.
- *Executed change order covers all damages* The contract should provide that the contractor is not entitled to any additional time or money for an event or circumstance that is the subject of a fully executed bilateral change order or a unilateral change order that has become binding.
- *Contractor required to keep working* The contractor should be required to keep working even if a change order proposal or other dispute is not resolved to the contractor's satisfaction.

### *Notice*

It is important that the owner receive timely notice when the contractor believes an event or circumstance entitles it to additional money and/or time. Timely notice allows the owner to mitigate its costs. The owner can minimize costs by declining to have the change order work done or by successfully arguing that it is not change order work (i.e., that it is base scope work already included in the contract). If the work must be done, costs can be controlled by electing the least costly method of doing the work and/or the most limited scope necessary.

Exhibit 9.1 illustrates two notice provisions. The first provision applies to proposals required by the owner, and the second applies to proposals submitted at the contractor's instigation.

The notice and proposal deadlines contained in the two provisions in Exhibit 9.1 are important tools for the owner in protecting its ability to mitigate the impacts of change orders and to efficiently manage the change order process. Some contracts combine the notice and proposal into a single document. The owner has two potentially conflicting interests which are not well served by combining the notice and the proposal. On the one hand, the owner needs to know as soon as possible when there is a potential change order. On the other hand, the owner wants change order proposals to be as complete as possible, and that often takes some time.

In order to satisfy both of these interests, it is recommended, in the case of potential change orders not requested by the owner, that the contractor be required to submit a notice very soon after the event or circumstance occurs that causes the contractor to believe a change order is warranted. This preserves the owner's mitigation options to the greatest extent possible.

The proposal itself, in order to be as complete as possible, will take some time to assemble. That is why it makes sense to give the contractor a reasonable period to develop the proposal. Neither side is well served by the owner rejecting a proposal because it is incomplete, and

---

**Exhibit 9.1   Change order notice provisions**

9.3 *Owner Requested Proposals* Owner may, in connection with any change in the Work made at its instruction, request a proposal from Contractor for the extra costs and/or time associated with the performance of such change, if any. Following the receipt of such a request, Contractor shall have twenty-one (21) days in which to submit its proposal, except that Owner may extend such period by written notice to Contractor. Failure by Contractor to submit its proposal within the twenty-one (21) days, or when applicable, the extended period of time, shall constitute a complete waiver by Contractor of any right to seek an increase in the Contract Price and/or an extension in the Contract Time associated with the subject of the invitation for proposal.

9.4 *Contractor's Notice and Proposal* Contractor shall, within five (5) days of the happening of an act, omission, or circumstance which it believes entitles it to an increase in the Contract Price and/or an increase in the Contract Time, submit a written notice to Owner of its intent to submit a change order proposal. Contractor shall have twenty-one (21) days from the date of its written notice to submit its change order proposal to Owner, except that Owner may extend such period by written notice to Contractor. Failure by Contractor to submit such notice within the five (5) days following the happening of such act, omission, or circumstance, or failure to submit such proposal within the twenty-one (21) days, or when applicable, the extended period of time, following submission of its notice shall constitute a complete waiver by Contractor of its right to seek any increase in the Contract Price or any extension of the Contract Time in connection with such act, omission, or circumstance.

---

particularly not by a series of rejections for incompleteness. For that reason, it is frequently appropriate for the owner to give time extensions to the contractor for completion of the proposal. On the other hand the owner cannot manage a project effectively if there are potential change orders outstanding for extended periods of time. It is important from the owner's standpoint to resolve change order proposals as quickly as reasonably possible. This is because unresolved change order proposals undercut effective control of scope, cost, and schedule. They are also seriously disliked by equity investors and lenders.

For the reasons indicated, it is important that the deadlines in the provisions in Exhibit 9.1 be met. That is why both provisions contain a statement that failure to comply with the deadlines will result in a waiver of the proposal. Compliance with these types of deadlines is an important goal for most contractors. Their efforts are generally assisted through a weekly change order meeting involving the owner and the contractor, and, often, the designer. This meeting, using a change order log prepared and updated by one of the parties, discusses the status of outstanding change order proposals. It is not used to evaluate the merit or negotiate the value of specific change order proposals.

There are contractors for whom internal procedures and the weekly change order procedures are not enough. In some cases, contractors will deliberately drag their feet on noticing potential change orders and/or submitting their proposals because they think it enhances their bargaining position. In these cases, the threat of waiving their ability to get additional compensation becomes a significant consideration. The owner should not be reluctant to enforce the waiver provision. However, to be legally sustainable, it must be enforced reasonably. Denying an otherwise meritorious proposal because it is one day late will not be upheld in court. On the other hand, a proposal submitted at the contractor's instigation after substantial completion of the project dealing with an issue that occurred eight months before and with no prior notice will likely be held to have been waived.

### Required documentation

It is important that the owner have sufficient documentation to analyze merit and, for proposals with merit, appropriate value. Even in cases where the owner requests a proposal or is knowledgeable about the event or circumstance prompting the contractor to submit a proposal, there should be documentation sufficient to support the proposal. Requiring documentation ensures that merit and value are appropriately substantiated and provides assurance to the owner that any growth in scope, cost, and/or schedule is controlled to the greatest extent possible. The owner must be able to require additional documentation if the contractor's original proposal is not sufficiently supported.

Exhibit 9.2 illustrates a provision that effectively addresses documentation for change order proposals.

---

**Exhibit 9.2   Change order proposal documentation**

9.6 *Sufficient Documentation* Each change order proposal submitted pursuant to Article 9 shall be submitted by Contractor to Owner and, at Owner's request with a copy to Designer, and shall be accompanied by documentation sufficient to substantiate Contractor's proposal. Proposals seeking an extension in the Contract Time shall include a schedule analysis substantiating the extension sought in the proposal. Owner may request additional documentation if in its judgment such additional documentation is reasonably necessary to determine if Contractor's proposal is justified as to entitlement and/or value.

---

### Unilateral change orders

The owner should be contractually authorized to issue unilateral change orders. A unilateral change order is a change order signed only by the owner. For those used to working with construction change directives ("CCD"), there will be some familiarity with this type of document. However there are two important differences. First, the unilateral change order only requires the signature of the owner. The CCD requires the signature of both the owner and the designer. Second, the CCD, if and when signed by the contractor, must become a change order. The unilateral change order is a change order from the time of issuance, and, therefore, requires no additional documentation if agreed to by the contractor. If the contractor agrees to the terms of the unilateral change order, it can either sign the change order, or simply not dispute it within the designated period, after which it becomes fully effective as if it were a bilateral (i.e., signed by both parties) change order.

The unilateral change order is a particularly useful mechanism when the owner and the contractor agree that a proposal has merit but cannot, after good faith negotiations, agree on the value. The unilateral change order allows the owner to create a resolution, or at least a potential resolution, by issuing the unilateral change order. Exhibit 9.3 illustrates a provision which authorizes unilateral change orders.

---

**Exhibit 9.3 Provision authorizing unilateral change orders**

9.2 *Written Change Orders Required* Changes in the Contract Price and/or the Contract Time shall only be made by written change order ("Change Order") pursuant to the terms and conditions of this Article 9. A Change Order shall become fully binding on Contractor when it is executed by both Owner and Contractor; provided that Owner may issue a Change Order executed only by Owner ("Unilateral Change Order") and such Unilateral Change Order shall become fully binding as a Change Order upon Contractor and Owner unless it files a notice of claim within five (5) days of receipt of such Unilateral Change Order and otherwise complies with Article 11.

---

It is necessary to allow the contractor to file a claim against a unilateral change order; otherwise no contractor would agree to the contract. The claim is in essence an appeal of the terms of the unilateral change order. However, even with the potential resort to the claims process, the unilateral change order can be an effective mechanism for breaking negotiating log jams. The claims process has its own set of deadlines which lead ultimately to resolution, so even if there is a claim, there will be a resolution of the issue.

### Damages for delay

If the contractor is delayed for reasons that are not the contractor's responsibility, it may be eligible for damages associated with that delay (usually called delay damages). Delay damages have two potential components. The first is an extension of time. The second is monetary compensation for additional costs. Change order proposals associated with delay will be discussed in some length later in this chapter. For drafting purposes, it is sufficient to say that, in order to reasonably protect the owner, the construction contract should provide that when the contractor is entitled to delay damages of any kind, those damages should be limited to an extension of time except for three circumstances. These are differing site conditions (which are totally out of the contractor's control), changes in the work requested by the owner which impact the schedule and create extra costs, and any other act or omission by the owner which directly causes the contractor to incur extra costs resulting from a delay. Exhibit 9.4 illustrates a provision which allows for contractor recovery under the circumstances just described.

---

**Exhibit 9.4 Provision on delay damages**

9.12 *Remedies for Delay* A time extension shall be Contractor's sole and exclusive remedy for any delay of whatever kind and however caused except for those delays directly caused by differing site conditions, changes in the Work ordered by Owner, or acts or omissions of Owner. Contractor shall be entitled to an increase in the Contract Price under this Paragraph only to the extent of increased costs directly caused by such differing site conditions, changes in the Work, or acts or omissions of Owner.

### *Executed change order covers all damages*

In order to prevent the contractor from revisiting issues which have been previously resolved, the contract should establish that the contractor is not entitled to any additional time or money for an event or circumstance that is the subject of a fully executed change order (or a unilateral change order that has not been disputed and therefore has gone into effect). The reason such a provision is important is because some contractors, if they are losing money on the job or are not making their projected profits, will look to make up financial ground by submitting more change order proposals. One potentially fertile ground is to revisit issues on which the owner has already acknowledged that it owed the contractor additional compensation. It is to forestall this type of revisiting of issues that the provision we are discussing is important. Exhibit 9.5 illustrates such a provision.

---

**Exhibit 9.5   Executed change order covers all damages**

9.13 *Change Order Full Compensation* The increase in Contract Price and/or extension in Contract Time included in any Change Order shall be Contractor's full compensation, including all direct, indirect, supplemental, and all other costs of any kind, for the acts, omissions, or circumstances giving rise to such Change Order, and Contractor shall not seek any additional compensation under Article 9 or Article 11 for such acts, omissions, or circumstances.

---

### *Contractor required to keep working*

The contractor should be required to keep working even if a change order proposal or claim is not resolved or resolved to the contractor's satisfaction. It is imperative not to give the contractor the leverage which would exist if the contractor could stop working if it didn't like the resolution, or lack of resolution, of a change order proposal or claim. Exhibit 9.6 illustrates a provision which requires the contractor to keep working regardless of the status of any of its change order proposals or claims.

---

**Exhibit 9.6   Contractor must keep working**

3.14 *Contractor to Keep Working* Contractor shall at all times prosecute the Work so as to complete the work by the contractually specified completion date. Contractor shall continue to perform the Work notwithstanding the existence of any outstanding change order proposals or claims, and Contractor shall not suspend or stop work for any other reason than a written instruction by Owner issued as provided by Paragraphs 7.6 or 7.7 or Article 12.

---

## Requirements for the contractor to prove change order proposals

This section explains what the contractor has to prove in order to be awarded extra costs and/or additional time. The first part of the section discusses the four elements the contractor has to prove for any change order proposal. The second part of the section summarizes what the contractor has to show, in addition to the four elements, to be awarded additional money and/ or time for a number of specific proposals.

There are four basic requirements for all change order proposals. The contractor must demonstrate actual extra costs and/or additional time, contractual entitlement, direct causation, and sufficient documentation.

### *Actual extra costs and/or additional time*

The purpose of a change order proposal is to allow the contractor to recover costs it could not anticipate incurring when it prepared its bid. Therefore, in order to be awarded an increase in contract price, the contractor must document that it actually will, or did, incur extra costs. Similarly, for an extension in contract time, the contractor must document that performing the work actually will, or did, take longer than shown in the currently approved schedule.

For example, suppose the contractor encountered a large boulder during excavation, and the boulder constituted a differing site condition under the contract. Suppose also that the contractor already had a large excavator and a large truck on site. In this case, excavating and removing the boulder may not result in extra costs. If there are no actual extra costs, there is no basis for a change order even though the boulder constituted a differing site condition.

### *Contractual entitlement*

To be awarded a change order, the contractor has to show that its position has merit; that it is "entitled" to the extra money and/or time. There are two aspects to determining entitlement: identifying applicable contractual provisions and determining how they apply to what actually happened.

The construction contract is the basis of the entire relationship between the owner and the contractor. Any claim for extra money and/or additional time must be based on the contract. The contractor must be able to point to one or more contractual provisions that serve as the basis for its proposal. A proposal based on extra work, for example, should be justified by a reference to applicable portions of the technical specifications. In this case, the contractor will be attempting to demonstrate that the work in question is extra work by showing it is not included in the current contract documents.

Contractors sometimes attempt to justify change order proposals on other bases. Three of the most common are summarized below.

- *Industry practice* The contractor asserts that the way it performed the work is the normal practice in the construction industry in the area of the project. Therefore, notwithstanding that the applicable contractual provisions contain different requirements the contractor is entitled to its extra costs.
- *Doing a good job* The contractor claims it is doing a good job for the owner and therefore, regardless of what the contract provides, it deserves additional money and/or time.
- *Losing money* The contractor alleges that it is losing money on the job and therefore, regardless of what the applicable contractual provisions provide, it should be awarded extra costs and/or additional time.

If there are no contractual provisions that support the contractor's position, none of these assertions should be accepted as substitutes.

The contractor's facts must support the theory of contractual entitlement that the contractor is offering. The contract may stipulate that the contractor shall receive additional time for days lost due to acts of God, but, if the contractor cannot show events constituting an act of God, it cannot justify additional time.

### *Direct causation*

The contractor must show that the extra costs and/or additional time it seeks were caused by the facts described in the proposal. The causal link must be direct, and not indirect or tangential.

For example, take a case where the contractor can demonstrate that it was interfered with by the owner. The owner instructed the contractor to have a portion of the plumbing work performed in a sequence different than planned by the contractor. There are extra costs associated with the re-sequencing of the plumbing work. However, the extra costs were not directly caused by the owner's interference because those extra costs actually arose from the contractor's failure to coordinate the plumbing and HVAC work. In this situation, the contractor cannot recover its extra plumbing costs.

### *Sufficient documentation*

The contractor's change order proposal must include sufficient documentation to substantiate the facts on which it bases its proposal and the extra costs and/or additional time it seeks. That is because the proposal must demonstrate that the contract and the actual events combine to create entitlement. The contractor must also document the costs and/or time it is requesting.

Four of the most common forms of documenting entitlement are correspondence, meeting minutes, photographs and/or videos, and the contractor's daily field reports. Exhibit 9.7 explains these types of documentation.

---

**Exhibit 9.7   Forms of documentation**

- *Correspondence (hard copy and electronic)* Copies of correspondence from the owner to the contractor can be used to show an instruction to the contractor or the owner's agreement with the contractor's positions. Copies of correspondence from the contractor to the owner can show the contractor explaining the facts involved in the claim. The owner's responding correspondence can be used to show the owner's acquiescence.
- *Meeting minutes* Minutes of meetings can also be used to evidence the owner's instructions, or the owner's acquiescence in the contractor's notice of how it intended to proceed.
- *Photographs and/or videos* The project's physical progress is recorded by photographs and/or videos. They can also document differing site conditions, adverse working conditions, and certain results of owner actions.
- *Contractor's daily field reports* These reports record which entities were performing what portions of the work on the date of the report. They will list which equipment was in the field, and what work each piece was performing. They identify the workers and supervisors present on that date. Field reports will show weather conditions and refer to any unexpected developments.

---

The contractor must also document the amount of extra costs it claims to have incurred. Costs can be substantiated in a number of ways. For labor costs, this includes copies of certified payrolls for the applicable periods. For material costs, this includes supplier invoices. Documentation of equipment costs, if owned, should include records of the number of days the piece of equipment was used, an explanation of the contractor's standard accounting practice for charging for the use of owned equipment, and the rate that standard practice has developed for the piece of equipment in question. For rented equipment, documentation should include invoices from the firm providing the equipment. For subcontractor costs, appropriate documentation includes subcontractor requisitions for payment and accompanying documentation. The owner should require the contractor to submit copies of cancelled checks or electronic payment when it questions costs.

The contractor must substantiate any requested additional time. The contractor must submit a schedule analysis comparing the scheduled time to the actual time for performing the work and explaining the variance.

### *Conclusion: a change order proposal must have all four elements*

Before awarding the contractor any extra money and/or additional time in response to a contractor's change order proposal, the owner must satisfy itself that the contractor's proposal includes all four of the elements just discussed. Exhibit 9.8 summarizes the requirements.

---

**Exhibit 9.8   Requirements for all change order proposals**

- A statement that the contractor actually will or did incur extra costs and/or actually will or did require additional time to perform the work;
- a showing that the contract provides for recovery given the facts set forth in the proposal;
- an explanation of how the events described in the proposal will or did directly cause the extra costs and/or additional time; and
- documentation sufficient to substantiate entitlement and the requested costs and/or time.

---

To effectively control scope, cost, and schedule, the owner must have an effective change order management program. A key element of that program is insisting that all change order proposals satisfy all four requirements. In addition, for the types of proposals discussed below, they must also meet the specific requirements for each type of proposal.

### What the contractor has to show for specific change order proposals

This section identifies 13 specific types of change order proposals. For each type of proposal, the section summarizes what the contractor has to substantiate in order to prove entitlement.

### *Acceleration*

Acceleration refers to situations where the contractor must use additional resources (i.e., labor, materials, equipment, and/or subcontractors) to perform work at a rate faster than provided by the currently approved schedule. Exhibit 9.9 shows the requirements for an acceleration proposal.

---

**Exhibit 9.9    Acceleration**

In order to recover, the contractor must show that:

- there was an excusable delay (i.e., a delay for which the contractor was not responsible);
- it was acting pursuant to an explicit or implicit order by the owner to accelerate, and not at its own initiative; and
- it actually did accelerate.

---

When analyzing an acceleration claim two points require attention: the contractor's obligation to recover schedule and the nature of the instruction to accelerate. Most construction contracts require the contractor, if it falls behind schedule for reasons for which the contractor is responsible, to use whatever resources are necessary to get back on schedule. Extra costs associated with this type of schedule recovery are the contractor's responsibility. In this type of situation, the contractor can show an instruction from the owner, acceleration, and associated extra costs. However, it cannot show excusable delay and, therefore, cannot recover.

There may be an issue as to whether any given communication really is an instruction to accelerate. Statements such as, "We're concerned about the extent to which you have fallen behind schedule," or "We would appreciate if you could finish [a certain portion of the work] early" would not be instructions to accelerate. A statement such as "We need [a certain portion of the work] by [the original date for completion] notwithstanding that the recent severe weather kept you from making scheduled progress, so please do whatever is necessary to complete the work by that date" would be an instruction to accelerate.

### Acts of God

An act of God, for purposes of change order proposals, is an event of nature that was unforeseeable. The contractor must satisfy both elements. It must demonstrate that the event is an act of nature and not done by a human or a human organization. A strike, for example, is not an act of God. Similarly, the event must not have been in the parties' contemplation when they entered into the contract; if it was, the contractor would be obligated to plan for it. The most typical example of an act of nature is severe weather. The contractor must show that the weather was unusually severe, both for the time of year, and for the area where the work is located.

When analyzing a severe weather proposal, the contract documents should be reviewed to determine if the contractor has been instructed to include in its schedule a specified number of days that will be lost due to severe weather. If there is such a provision in the contract, then the next question is whether the contractor has used all of those days. If not, this proposal should be rejected, at least to the extent of the remaining unused days.

### Cardinal change

This type of proposal asserts that because of actions taken by the owner, the agreement between the parties has been changed to an extent not contemplated by the parties when they entered into their agreement. Exhibit 9.10 shows the requirements for a cardinal change proposal.

---

**Exhibit 9.10   Cardinal changes**

In order to recover, the contractor must show that:

- as a result of actions taken by the owner, the nature of the work is substantially changed (i.e., altered beyond the typical changes envisioned by the change order provisions in the contract); and
- the resulting costs incurred by the contractor are not adequately addressed by the contract's change order provisions.

---

This type of proposal often alleges the cumulative effect of a large number of change orders that individually were appropriately processed through the contract's change order provisions. Whether it is a single change order, or the cumulative effects of a number of change orders, the essence of this proposal is that the owner has changed the agreement between the parties, and, therefore, the contractor is entitled to recover costs to which the contract does not otherwise entitle it.

### Defective specifications

The contractor must show that the work in question had to be performed because of a deficiency in the plans and/or specifications. The most common forms of this claim assert that the plans and/or specifications are incomplete, ambiguous, inconsistent, or do not meet building code or other legal requirements.

The contractor has to show that the design documents were in fact incomplete, ambiguous, inconsistent, did not meet legal requirements, or were otherwise deficient, and that as a result, the contractor will or did incur extra costs in performing the work. These extra costs result from additional scope being required to resolve the deficiency. The additional scope may be in the form of extra work not currently shown or it may take the form of doing the same work in a more complicated way than reasonably anticipated from the drawings.

If there is a contractual requirement that the contractor review the documents prior to commencing the work to identify any problems, the contractor, in order to recover, must show that:

- it reviewed the documents carefully; and
- notwithstanding its careful review, it reasonably was unable to discover the problem.

If the contractor finds a problem during this initial review, it has an obligation to inform the owner and request a resolution. Generally, if problems are discovered prior to commencing work, they can be resolved at much less cost to the owner than if they have to be resolved during construction.

There is a distinction in this type of situation between a patent and a latent problem. If the problem was or should have been obvious to the contractor when it reviewed the documents, a patent problem, then the contractor will not be able to show its failure to discover the problem was reasonable. If the problem is difficult to discover, a latent problem, and the contractor has undertaken a reasonable review, it will not be denied recovery because it did not find the problem.

### *Delay*

Any change order proposal that says that a portion of the work will, or did, take longer than allowed for that part of the work on the currently approved schedule, and requesting an extension of time, is a delay proposal. Such a proposal may also request an increase in the contract price as a result of the delay.

Examples of delay proposals asserted by the contractor include those alleged to result from the owner ordering changes to the work; interfering with the contractor's performance of the work; suspending the work; and taking longer than allowed by the contract to issue approvals/disapprovals, contract interpretations, and other required actions.

To sustain a delay proposal, the contractor is required to show that it was delayed through no fault of its own. That is, one or more construction activities will or did take longer to perform than shown on the currently approved schedule for reasons that are not the contractor's responsibility. Today, most construction contracts require the contractor to use a CPM schedule. Generally, contracts requiring the use of CPM schedules provide that the contractor has to show that all available float has been utilized and that the delay in question has impacted the critical path. Exhibit 9.11 summarizes what the contractor has to show for a delay proposal, assuming a CPM schedule is required.

---

**Exhibit 9.11    Delay proposals**

The contractor has to show that:

- it was delayed through no fault of its own;
- all available float has been utilized; and
- the delay in question has impacted the critical path.

When analyzing a delay claim, the three key questions the owner should consider are:

- was there in fact a delay;
- if so, whose responsibility was it; and
- the duration of the delay for which contractor is not responsible.

---

The principal tool in addressing those questions is the project schedule. For those not familiar with schedules and scheduling we will briefly detour to explain some fundamentals of scheduling. The intent is only to assist the person reviewing the delay proposal on behalf of the owner to ask the right questions, not to become a scheduler.

### *Function of construction schedules*

Construction schedules have three principal functions. The first is contractor planning. The project schedule is the mechanism the contractor uses to plan in what order it is going to perform the work.

The second function is project monitoring. The owner and the contractor use the schedule to determine if the project is going to be completed on time. This is accomplished by comparing the physical progress of the work against the current schedule. The current schedule is the one most recently submitted by the contractor and approved by the owner.

The third function of schedules is to analyze delay proposals and claims. The schedule is used to determine if the contractor's request for additional time has merit, and, if so, how much additional time should be granted.

*Scheduling terminology*

Before getting into the discussion of schedules, it is useful to define several frequently used scheduling terms. These terms include activity, duration, start and end dates, logic, predecessor and successor activities, constraints, critical path, and float.

- *Activity* An activity is one element of the work. On a simple bar chart schedule, an activity might be completing the foundation. On a more detailed schedule, excavating, erecting forms, placing concrete, finishing the concrete, and removing the forms will be separate activities.
- *Duration* The length of time it will or did take to perform an activity starting on its start date and ending on its end date. The use of the words "will or did" reflects the reality of schedules during the life of a project. The original schedule, often called a baseline schedule, is entirely a projected schedule. It is the contractor's best guess as to how the project will be performed. So, at the beginning of the project, each activity is in the "will take" category. However, most construction contracts require monthly updates of the schedule and further require that the updates reflect progress in the field. So, as the project progresses, certain activities on the schedule have actually been completed and their duration is in the "did take" category.
- *Start and end dates* Each activity starts on a particular date and ends on a particular date. The date it will or did start is the start date and the date it will be or was completed is the end date. On a CPM schedule there is usually an early start date and a late start date. The early start date is the earliest date on which the activity can be started, given the work that must be completed before that activity can begin. The late start date is the latest date the activity can start without delaying the completion of the project, given the activities that must follow to complete the project. Similarly, there is an early end date and a late end date. The early end date is the earliest date the activity can be completed given its early start date and its duration. The late end date is the latest date on which the activity can be completed without delaying the project. Early and late start and end dates are only applicable to projected activities. Activities that have been completed have an actual start date and an actual end date.
- *Logic* The logic of a schedule refers to the order in which activities are to be performed. It is the heart of the scheduling process. The contractor determines the most efficient approach to building the project. Certain relationships are required by the laws of science. A building's foundation must be built before its structure can be built. However, many choices are left to the contractor's discretion. For example, the contractor will consider whether it is more efficient to close in a multi-story building from the top down, the bottom floor up, or start in the middle and work in both directions. At the activity level, logic refers to those activities that must be completed before others can be started. In fact, there are more relationships. Certain activities must wait for another activity to start

before that activity can start, but it is not necessary that the earlier activity be fully completed before the later activity starts. For example, on any floor of the multi-story building, the structural work must be completed before exterior cladding can be installed. However, on that same floor, installation of the drywall has to be started before the painting can start, but the drywall does not have to be completed in order for painting on that floor to start.

- *Predecessor and successor activities* An activity which must be completed before another activity can start is the predecessor activity. The activity that can only be performed after the earlier activity is completed is the successor activity.
- *Constraints* Constraints are events or circumstances that impact the logic. Some constraints are physical, as in the requirement that the foundation be placed before the structural elements can be erected. Others are business related. For example, there may be a shortage of electricians in the project area labor market. That shortage may translate into fewer electricians available to work on the project than the electrical contractor could actually use. This, in turn, will mean the work cannot proceed as quickly as if there were more electricians available. Other constraints are governmental. Frequently, municipal ordinances will dictate when work on a project can start and by when it must end each weekday and on weekends and holidays. These requirements may impact how fast the work can be performed.
- *Critical path* The critical path consists of those activities which, if any one of them is delayed, the project's completion date will be delayed. As the construction of the project progresses, the activities on the critical path can and frequently do change.
- *Float* While there are several types of float, the basic point is that float is the difference between the date when the activity is planned to be finished and the date when it must be finished so as not to delay the next activity. For example, consider a project consisting of the construction of a 20,000 square foot office building, which is scheduled to be completed in 360 days. If installation of the roof is scheduled to be completed on day 140, but the installation could be completed on day 146 without delaying the completion of the project, there are 6 days of float.

*Two most common types of schedules*

The two most common schedules are the bar chart and CPM schedules. Exhibit 9.12 lists the information shown by both types of schedules.

---

**Exhibit 9.12    Information on bar chart and CPM schedules**

Both schedules show the following types of information:

- the activities involved in building the project;
- the length of time each activity will take;
- when the activity will start and when it will end;
- which activities have to be completed before which other activities can be started; and
- when the project will be completed.

The principal difference between a bar chart schedule and a CPM schedule is the level of detail involved in assembling and presenting the schedule.

BAR CHART SCHEDULE

A bar chart schedule identifies a list of activities required to perform the work down the left side of the chart. Across the top of the chart is the contract term broken down weekly or monthly. Next to each activity, the contractor draws a bar showing when the activity will start and when it will end. The length of the bar represents the activity's duration. Bar chart schedules usually show a relatively limited number of activities, and depict predecessor–successor relationships by implication; that is, by the position of the various activity bars in relation to each other.

CPM SCHEDULE

A CPM schedule typically includes many activities. The schedule is represented in two forms. One is a network diagram showing all the activities and how they relate to each other. This diagram connects each project activity to those which must be completed before it can be started (predecessor activities). The diagram also connects each activity to those which cannot begin until it has been completed (successor activities). The second form is a printout containing a table showing each activity's early start and finish dates and a bar chart depicting these start and finish dates and the resulting durations.

The principal reason for using a CPM schedule is that it shows the critical path. The activities that are on the critical path change as the work progresses. For example, at the beginning of a project involving the construction of a bridge, painting the bridge may not be on the critical path because it can be done concurrently with landscaping and placing asphalt. However, because the hoist used by the bridge painters is unavailable, and painting is delayed, it becomes a critical path activity because all the activities planned to be concurrent are now completed and landscaping, which is all that is left other than painting, cannot be completed near the bridge until the painting is completed.

*Understanding float*

The significance of float has to do with the provision in most construction contracts that require a CPM schedule that the contractor is only entitled to additional time if the delay in question extends the critical path. That, in turn, depends on the existence and ownership of float. The contract should provide that neither party owns the float, and that it is to be used for the benefit of the project. This means that the contractor cannot bank float for its preferred use later in the project and must use available float before an activity can be considered delayed. The contractor is not entitled to any additional time in connection with a delayed activity unless the float is exhausted.

For example, the installation of window frames in an office building is scheduled to take 25 days, and the next activity, installation of the windows, can begin as many as 32 days after the start of the window frames before it impacts the critical path. This means there are seven days of float. If the contractor is delayed in window frame installation for three days for reasons that are the responsibility of the owner, the contractor is still not entitled to any additional time because three of the seven available days of float can be used, thereby avoiding an impact to the critical path.

*Using the schedule to analyze delay claims*

The schedule is the primary tool in analyzing delay proposals. There are three steps to analyzing a delay proposal. Exhibit 9.13 describes the three steps.

---

**Exhibit 9.13    Steps to analyzing a delay claim**

1    *Analyze schedule to date of delay* The schedule is examined to determine whether the contractor is behind schedule prior to the events that serve as the basis of the proposal.
2    *Analyze schedule for period of proposal* The schedule is analyzed in terms of the events that are the basis of the proposal to determine their impact.
3    *Determine responsibility and additional time* This step involves analyzing more than just the schedule. However, the schedule may indicate responsibility by revealing which activities caused the delay. This analysis, in combination with other documentation, will either confirm or contradict the contractor's claim of owner responsibility. If the contractor is entitled to some additional time, the analysis of the schedule will be an important indicator of how much time the contractor should get.

---

The delay proposal may also request compensation for extra costs. How to calculate those costs, if the contractor is entitled to them, will be discussed further on in this chapter.

### Denial of access to the site

The contractor must show that the owner is contractually obligated to provide it access to the site, and that the owner failed to provide access as required by the contract. Since the owner by definition owns or leases the site, it is always obligated to give access to the contractor. Therefore, if the contractor cannot perform its work because it has been denied access, it has met its entitlement standard for this type of claim.

A more complicated situation exists if the contract requires phased access. An example is the rehabilitation of a school while the school remains in operation. In the first phase, the students are relocated from the East Wing to allow the contractor access. When the East Wing is completed, the students are relocated there from the North Wing. The contractor then completes the North Wing. In this type of situation, failing to allow the contractor access to a wing when the contractor was scheduled to begin work in that wing would constitute denial of access to the contractor.

In this situation, the contractor's ability to show damage will depend on whether it did perform, or could reasonably have performed, other work on the project. If so, the denial of access may not have caused extra costs and/or a need for additional time.

### Differing site conditions

A differing site condition involves conditions that differ materially from those described in the contract. In some construction contracts, such as those issued by the federal government, a differing site condition also includes conditions that differ materially from those normally expected when performing the type of work required by the contract. Exhibit 9.14 indicates what a contractor has to show to substantiate a differing site condition proposal.

**Exhibit 9.14   Differing site conditions**

To prove a differing site condition, the contractor must show that:

- the condition is a physical condition at the site of the work;
- the condition differs significantly from the contract description (or the normal expectation, if that is included in the contract);
- the contractor has provided the contractually required prompt notice to the owner; and
- the contractor has satisfied any contractual requirement for a site investigation.

When analyzing proposals alleging differing site conditions, it is important to focus on each of these requirements.

- *Physical conditions at the site* The condition must relate to physical conditions at the site. An unexpected change in circumstances not related to a physical condition is not a differing site condition. A strike or an unanticipated shortage of materials is not a differing site condition.
- *Differs significantly* It is important to recognize that this standard has two parts, "differs" and "significantly." The claimed condition must actually be different than what is shown on the plans and/or described in the specifications (or than what is reasonably anticipated given the nature of the work, if that is a contractual standard). And the difference must be significant; a minor discrepancy does not meet the standard. What constitutes a significant difference depends on the nature of the project. For example, a large piece of concrete not shown on the plans, discovered while excavating for a new school building in an open suburban site, may well be a differing site condition. A similar piece of concrete, also not shown on the plans, may not be a differing site condition if encountered as part of demolishing and replacing an office building in the middle of an older city (particularly if the contract documents warn the contractor about potential debris on the site because of the site's history).
- *Contractually required prompt notice* Prompt notice to the owner is important. A differing site condition once discovered must be dealt with as soon as possible or physical problems at the site may increase rapidly.
- *Site investigation* If the contractor was required to visit the site, and if such a visit would have disclosed the condition to a reasonable contractor, then failure to undertake the required site visit precludes recovery by the contractor.

### Extra work

If the owner requests the contractor to perform work which is not part of the scope of work as described by the most current edition of the plans and specifications, the contractor's proposal is for extra work. This type of proposal is frequently solicited by the owner when it wants to make a change to the project. Sometimes the owner and the contractor do not agree on whether a particular request or instruction by the owner constitutes extra work. That type of situation is sometimes called a constructive change order. Exhibit 9.15 shows what the contractor must show to recover on an extra work proposal.

---

**Exhibit 9.15   Extra work**

The contractor is entitled to recover when it can prove that the work in question was:

- in fact extra work (that is, the work is not depicted or described in the most current version of the plans and specifications); and
- performed at the instruction of the owner.

---

Even if the work was extra, but it was not performed at the owner's direction, the contractor cannot recover. This is because if the contractor was entitled to be paid for whatever extra work the contractor elected to perform, the owner would lose control of the scope of work and the cost of the project.

A constructive change order can occur even after a bilateral change order has been executed. This is because the owner may order further work on the same matter. The contractor contends the work is extra, while the owner argues that the work is original scope work, or is work included in the previously executed bilateral change order.

### *Improperly denied substitution*

Most construction contracts allow the contractor to propose substitutions. Contractors will often have valuable suggestions concerning alternative materials and/or equipment. Contractors also propose substitutions to lower costs. Under a lump sum contract, if the contractor can lower its costs, it increases its profit. To assist the owner to protect scope, cost, and schedule, most construction contracts establish criteria for acceptable substitutions. These typically include requirements that the proposed substitution:

- be of equivalent or higher quality;
- provides the same or better functional result; and
- does not raise the cost of the project.

The most effective control over substitutions by the owner is provided by a provision that says the owner can accept or reject proposals in its sole discretion. Such a provision is illustrated in Exhibit 8.5.

Many public, and some private, construction contracts give the contractor the right to propose substitutions and require the owner to approve them if they meet certain criteria (typically the ones stated above). If the owner under such a contract inappropriately rejects a proposed substitution, causing the contractor to incur higher costs, the contractor may have a meritorious change order proposal. Exhibit 9.16 shows what the contractor must demonstrate to recover for improper denial of a proposed substitution.

### *Interference*

Interference proposals, also referred to as "disruption" or "hindrance" proposals, involve action by the owner, not contractually authorized, which forces the contractor to perform the work using a different method and/or in a different sequence than it had planned. Exhibit 9.17 lists the requirements for this type of proposal.

---

**Exhibit 9.16    Improper denial of substitution**

The contractor must show that:

- there is a contractual right to propose substitutions for the materials and equipment specified in the contract;
- it met the procedural and substantive requirements of the contract relating to substitution, if any; and
- nonetheless, the owner denied the substitution.

---

---

**Exhibit 9.17    Interference**

The contractor must show that:

- the owner took some specific action;
- the action was not contractually authorized; and
- that action caused the contractor to have to change its method and/or sequence for performing the work.

---

When analyzing interference proposals, it is important to remember that contractors frequently elect, at their own initiative, to change their techniques and sequences for performing the work. Those decisions are made based on the contractor's expertise, and often benefit the project and/or enhance the contractor's profit, both of which are appropriate. These changes are made at the contractor's instigation, and, as such, cannot serve as the basis for an interference claim.

For an example of interference, consider a project involving the construction of an office building. The schedule calls for the lobby to be the last portion of the project to be completed. When the project is about half complete, the owner requests the contractor to complete the lobby within three weeks because the owner thinks having a completed lobby will facilitate the leasing of the building. This change involves no extra work, but it does involve re-sequencing the work with likely additional costs. These costs might include extra labor costs for more workers and/or an additional shift in order to meet the compressed schedule and the costs, if incurred, would be recoverable from the owner.

*Performance impossible or impractical*

The contractor must show that the contractually required performance is either physically impossible, or so expensive relative to the original intent of the parties as to make it commercially impractical. An example involves the restoration of a historically significant building. The contract calls for certain historically accurate door hardware. However, when the contractor is ready to purchase this hardware, the contractor discovers that it is no longer manufactured as a standard item. The extra costs associated with identifying, purchasing, waiting for, and installing custom-made replica hardware are the owner's responsibility.

*Problems caused by other contractors*

Certain public owners are required to have multiple prime contracts on the same project. Other owners, both public and private, elect for various reasons to have more than one prime contractor working on a single project. Under these circumstances, one prime contractor can impact the work of another, causing the latter contractor to submit a change order proposal to the owner. Exhibit 9.18 shows what the contractor has to demonstrate to recover on a proposal based on problems caused by another contractor.

---

**Exhibit 9.18   Problems caused by other contractors**

The contractor must show that:

- its work was damaged, delayed, or interfered with;
- the damage, delay, or interference was caused by the actions taken by another contractor; and
- the other contractor was performing work for the same owner.

---

The most effective way for the owner to mitigate this type of proposal on projects with multiple prime contracts is to insert a provision in each contract providing that extra costs caused by another contractor are to be resolved directly between the contractors and is the responsibility of the contractor causing the damage.

*Wrongful termination*

A proposal alleging wrongful termination must show that the owner did not in fact have cause to terminate the contract. The contractor's position is that the owner failed to demonstrate that whatever grounds the owner invoked as a basis for termination were justified. Therefore, the substance of the contractor's proposal is dependent on the grounds the owner used to terminate the contract.

As an example, the owner may have based its termination on the ground that the contractor performed work that did not comply with the contract. The contractor might assert one or more of the following.

- The owner did not comply with the procedural requirements of the contract because it sent the notice actually terminating the contract before the end of the cure period.
- The work in question complies with all applicable contractual requirements, and therefore is not defective.
- The work in question was inspected by the owner and accepted; therefore it cannot now be labeled as defective work.
- The work in question substantially complies with applicable contractual requirements and the minor discrepancies that the contractor acknowledges cannot serve as grounds for termination under a contractual provision that specifies "material" deviations as the basis for termination.

## Common owner defenses

This section discusses seven of the most common defenses available to the owner in response to proposals submitted by the contractor. Each of these defenses is potentially applicable to one or more of the change order proposals discussed in the previous section.

### *Change order includes all costs*

The contract should stipulate that when a change order executed by both parties awards extra costs and/or time, the contractor is precluded from obtaining any additional money and/or time for the circumstances covered by that change order. The contractor may be inclined to revisit the facts that gave rise to a change order in an effort to obtain additional compensation. The owner will have to show that the facts underlying the current request for additional costs are the same facts that were the basis of the previously executed change order.

### *Defective work*

The owner is not required to accept or to pay for defective work. Defective work is defined as work that does not comply with applicable contractual requirements. The owner must show that the work in question did not comply with the applicable plans, specifications, and/or other contractual provisions. A frequent use of this defense is in response to work claimed by the contractor as extra work. If the work in question involves replacing defective work, it is not extra work.

### *Designer's decision is final*

If the contract says the decision of the designer on some or all matters is final and binding, the designer's decision will be accepted by the courts as final and binding. The language must evidence a clear intent to grant the designer that much authority. For that reason, the words used must be "final and binding," or words with equivalent meaning. If this language appears in the contract, the only basis for attacking the designer's decisions covered by the language is to show that the decision was arbitrary or made in bad faith.

### *Failure to comply with procedural requirements of the contract*

Most jurisdictions accord significant weight to the procedural requirements of a construction contract. If the contractor fails to comply with the procedural requirements of a construction contract, it may be precluded from recovery, even if it could otherwise demonstrate real loss. In some jurisdictions, the contractor's failure to comply with procedural requirements will only preclude the contractor's recovery if its failure to comply caused prejudice to the owner.

There are important policy considerations behind these requirements. It is important for owners, public and private, to control and account for project costs. The contractual requirement for timely notice provides the owner with two options.

- *Avoid the costs* The owner may avoid the costs by electing not to incur them or by successfully arguing that they are not the owner's responsibility.
- *Control the costs* The owner can seek to control costs by selecting lower cost alternatives or electing to only have work performed that is absolutely necessary.

Proper notice equips the owner to account for potential additional expenditures in the most appropriate manner. This is an important consideration for owners as they track the project's actual and potential cost against the project's budget.

### Lack of authority

The contractor may seek compensation based on a work directive from a member of the owner's team. The owner will be able to defeat this type of proposal if it can show that the member of the team did not have authority to issue the directive.

The authority to give instructions on a construction project may be either real or apparent. Real authority is where the person issuing the instruction to the contractor in fact has the authority to issue it. Apparent authority is where the person does not, in fact, have the authority to issue the instruction but reasonably appears to the contractor as having such authority. If the contractor reasonably relied on the authority of the person giving the instruction, the owner is responsible for the extra costs and/or time resulting from the instruction.

There are three ways to mitigate the authority problem. The first is to specify contractually which individuals have authority to bind the owner and to state that no other persons have authority to bind the owner. The second is to limit such authority to specified individuals through internal written procedures. The third is to specifically identify who has authority at the first progress meeting and to ensure it is accurately reported in the meeting minutes.

### Lack of causation

A previous section of this chapter described the need for the contractor to demonstrate causation. If the contractor cannot prove a direct link between the facts that it asserts and the extra costs and/or time which it claims to have incurred, the contractor cannot recover. Therefore, the owner can prevail if it can demonstrate that even if everything the contractor alleged is factually accurate, these events did not cause the contractor to experience extra costs or to take extra time.

For example, a proposal for extra work may include the project manager's salary for the four days of additional time the extra work involved. However, the owner can demonstrate that the project manager was required on the site for two additional weeks (including the four days) to ensure satisfactory completion of the landscaping work, the timing of which was not the owner's responsibility. Under these circumstances, the project manager's salary is not the owner's responsibility.

### No damage for delay

No damage for delay clauses provide that if the contractor is delayed, its only recourse is an extension of the contract time. It cannot get extra costs. The strictest form of this provision states that the contractor cannot recover extra costs associated with a delay regardless of which party caused the delay. A commonly used version provides that the contractor can only recover money damages for delay if the owner specifically causes the delay; otherwise, the contractor's only recourse is additional time.

## Calculating damages

This section discusses how the value of change order proposals should be calculated. It starts by reviewing the purpose of compensation. It will establish that compensation is based on what the contract allows. It then reviews the various categories of costs that are allowable in valuing a change order proposal, and the various approaches used by contractors to calculate change order value. Finally, it discusses how the owner should calculate the value of a change order.

The purpose of compensation (additional money and/or time) is to place the contractor in the same position as if the circumstances forming the basis of the change order proposal had not occurred. This means that each part of the contractor's claimed compensation must meet the "but for" test. The compensation recoverable by the contractor are the extra costs and/or additional time that would not have been incurred "but for" the circumstances described in the proposal. The contractor should not be able to recover costs for which it is responsible just because the changes have been cleverly included in a proposal for costs which it is entitled to recover.

The starting point for analyzing the compensation requested by the contractor is reviewing the contract. The contractor is only legally allowed the compensation that is authorized by the contract. This relates to the value of the extra costs and/or the number of additional days to be awarded to the contractor as much as it relates to the threshold question of whether the contractor is entitled to any compensation. Construction contracts specify how compensation is to be calculated, and which costs are allowed and which are not. These provisions should be thoroughly reviewed before analyzing any change order proposal submitted by the contractor.

There are seven categories of costs for which the contractor can seek recovery. Exhibit 9.19 lists these categories.

---

**Exhibit 9.19   Change order cost categories**

The categories of potential costs that can be claimed in change orders include:

- labor costs;
- material costs;
- equipment costs;
- subcontractor costs;
- contractually specified mark-ups;
- supplementary costs; and
- home office overhead costs.

---

The first four categories are all direct costs. They are costs directly incurred in the performance of the work. The overhead and profit mark-up is designed to compensate the contractor for overhead costs, and for its right to make a profit on its work. Supplemental costs and home office overhead costs are indirect costs, to which the contractor may, depending on the terms of the contract, be entitled as a result of the events and circumstances which justified the change order. It is important not to award indirect costs in direct cost categories, since the owner will then be paying for those costs twice. Each of the seven cost categories are discussed below.

### *Labor costs*

Direct labor costs are the costs associated with those persons actually performing the work. This includes trade workers and working foremen. The costs include the wages paid together with the labor burden, which includes the cost of health and welfare and other benefits and the contractor's share of the cost of the government mandated programs: social security, workers compensation, and unemployment compensation.

The labor burden is expressed either in terms of actual costs or as a percentage rate that is then applied to the total of the wages included in the change order proposal to calculate the cost of the burden. If expressed as a percentage rate, the rate is derived by dividing the total burden cost for a base period, usually performance of a portion of the original scope of work, by the total cost of wages for the same period.

Which level of supervision is included in direct labor costs is an important issue. A first line supervisor who is actively involved in the performance of the work, usually called a working foreman, is included in direct labor costs. A supervisor whose function is to coordinate the work of various trades and/or subcontractors, usually called a general foreman or superintendent, is not included in direct labor costs. Their labor costs are paid for from the funds generated by the mark-up for overhead.

In certain situations, the claimed labor costs will include costs associated with the contractor's loss of productivity. These costs arise from the workers not being able to perform the work at the standard rate at which they were anticipated to perform the work.

Productivity is measured in units of production. For example, a brick layer will lay so many bricks per hour. A painter will paint so many square feet per hour. To demonstrate loss of productivity, the contractor must show that the applicable unit of production is lower than it should have been. What the rate should have been is normally demonstrated in one of four ways which are described in Exhibit 9.20.

---

**Exhibit 9.20   Methods of measuring productivity**

- *Previous experience* The contractor compares the actual productivity rate on the current project to its experience on similar work on other projects.
- *Measured mile* The contractor compares the actual productivity rate on the current project to its experience on another portion of the same project.
- *Industry standards* There are materials published annually which contain productivity rates for the various construction trades. The contractor compares the production rate in the claim with the comparable rates in an appropriate publication.
- *Expert opinion* The contractor retains an expert to state what the productivity rate should have been. The contractor compares the actual production rate on the current project to the expert's opinion.

---

Among the most common causes of lost productivity are delays in delivery of supplies, re-sequenced work at owner's or contractor's instigation, coordination problems, limited access to the work, severe weather or other physically difficult working conditions, and poor supervision. Certain of these causes are the responsibility of the owner; others are the responsibility of the contractor. For that reason, a claimed loss of productivity, even a documented loss, does not automatically mean that the owner is responsible for the associated costs.

### Material costs

There are two categories of material costs: materials and installed equipment. Material costs include the costs of all materials installed as part of the work that is the subject of the proposed change order. Material costs include the full cost of the materials as paid by the contractor, which usually includes the cost of transportation and storage.

Most projects require the installation of equipment. If installed equipment is part of a change order, the costs would include the cost of the equipment and the costs of transportation and storage. In many instances, this type of equipment requires field service by the manufacturer as part of installation, start-up, and/or testing. Field service costs, if applicable, would also be included.

### Equipment costs

These costs are for the use of equipment used to perform the work. If the equipment is rented, the allowable cost is the cost of rental for the period covered by the proposal, plus any operating costs paid by the contractor.

If the equipment is owned, then the allowable cost is the cost of ownership for the period covered by the proposal. These costs appropriately include such elements as:

*   amounts paid for acquisition of the equipment, including financing;
*   fuel;
*   maintenance;
*   storage; and
*   depreciation.

To recover ownership costs, the contractor must be able to show that the costs included in the proposal are reasonable. One way to do that is to establish a standard accounting procedure for identifying and quantifying equipment costs and to demonstrate that the contractor includes the costs produced by that accounting procedure in its estimates. Another common approach is to establish a separate company which owns the equipment and from which the contractor rents the equipment it uses. Under these circumstances, the contractor simply has to demonstrate that the rental rates it pays its corporate affiliate are reasonable.

The third type of equipment cost is the small tools that are used in the performance of the work. These tools are generally consumable, meaning they get broken, lost, or otherwise are in a condition that they cannot be used again. Some contracts allow these tools as a direct cost. Because they are difficult to quantify accurately, it is recommended that they be dealt with in one of two ways. The contract should specify that the costs of small tools is covered by the mark-up for overhead, or it should allow the contractor to charge a set percentage (usually 1 or 2 percent) of the direct labor costs claimed in the proposal to cover the cost of small tools.

### Subcontractor costs

Change order proposals almost always involve work performed by subcontractors. Costs paid by the contractor to these subcontractors are allowable in calculating the value of the change order.

### Contractually specified mark-ups

The contract usually authorizes the contractor to mark-up its costs that are allowed in change orders and claims. There are several mark-ups. The three most common are overhead and profit, costs of insurance premiums, and costs of bond premiums.

Overhead and profit are usually a combined mark-up, and it is applied to the total of the direct costs, labor, materials, equipment, and subcontractors. It is designed to compensate the contractor for overhead costs and to assure that it makes a profit on its change order work. In today's construction environment in which general contractors and construction managers usually do not self perform work, the mark-up is divided between the subcontractor that performs the work and the contractor. For example, the construction contract may provide that the subcontractor is allowed a 15 percent mark-up for overhead and profit and the contractor is entitled to 5 percent. The contractor's share of the mark-up is to compensate for management costs. In this example, the contract should provide that the maximum mark-up on any change order is 20 percent. If there are second and lower tier subcontractors, the entity performing the work will be entitled to the 15 percent mark-up, and the higher tier entities split the remaining 5 percent as determined by the contractor.

The contract requires the contractor to fully insure its work, and the premium increases as the value of the work being insured increases. Therefore, the contract allows the contractor a mark-up for insurance. This mark-up is applied to the total of the direct costs plus the mark-up for overhead and profit; that is, it is not included in the total to which the overhead and profit mark-up is applied.

If the contract requires the contractor to obtain payment and/or performance bonds the contract will allow the contractor a mark-up in each change order for the increase in bond premiums associated with the increased value being covered by the bonds. This mark-up is applied to the total cost of the proposal including the insurance mark-up.

In construction manager at risk contracts mark-ups are treated somewhat differently. As discussed in Chapter Seven, overhead costs are detailed in a part of the contract called general conditions costs. Change orders under a construction manager at risk contract have two mark-ups: one for general conditions costs (i.e., overhead) and one for the CM fee (i.e., profit). The mark-ups for insurance and bond premiums are calculated as described above.

## *Supplemental costs*

Supplemental costs are additional overhead costs actually incurred by the contractor as a result of performing the work covered by the change order proposal. They would include such things as additional supervision, telephone calls, office supplies, and additional trailer space necessitated by the work covered by the proposal.

There is a distinction between overhead costs that are includable in supplemental costs and those that are not. Overhead costs that are directly attributable to the work in the proposal are includable in supplemental costs. If an additional general foreman is required to manage the proposal work, or the original general foreman has to stay on the job for some additional period of time (for reasons that are not the contractor's responsibility) the resulting labor costs would be part of supplemental costs.

Overhead costs that support the entire project are not includable as supplemental costs. Examples include the costs of the site trailer, the acquisition of the phone system, and the on-site computers. Those costs are included in the original bid and are further reimbursed through the contractually specified mark-up for overhead and profit. However, each of these costs could be part of supplemental costs if they are incurred for an additional period of time (for which the contractor is not responsible).

The contractor can justify recovery of supplemental costs in two instances. The first is when it can demonstrate that its overhead costs are greater than bid for reasons that are the responsibility of the owner. These costs would be additional supplemental costs. The second instance is to reimburse the contractor for overhead costs incurred during the additional time it spent on the project as a result of delays for which it was not responsible. These costs are called extended supplemental costs or extended overhead.

### *Home office overhead costs*

Home office overhead is based on the concept that the contractor has a home office that provides a number of functions which support the performance of work on each of its projects. These functions, some of which were listed in the discussion of supplemental costs, are overhead with respect to individual projects. Because these costs are overhead, the only way they can be reimbursed is through contributions from each of the projects performed by the contractor.

If a contract is suspended for any significant period of time, the revenues from the suspended contract are not available to contribute to the home office overhead costs. These are fixed costs that do not change when one contract or another is started, suspended, or completed. Under contracts that recognize home office overhead as a recoverable cost, the owner may be responsible for reimbursing missing contributions to home office overhead if it is responsible for interrupting those contributions.

Typically, the owner is held responsible for interrupting contributions to home office overhead if the owner suspends the work and requires the contractor to standby on site for a significant period of time. The owner has the burden of proving that the contractor could have obtained additional work during the period of suspension, which would have resulted in some contribution to home office overhead, but, for reasons within the contractor's control, it did not obtain such work.

Home office overhead costs are calculated by use of a formula that converts the costs to a per day amount. The per day amount is then multiplied times the number of days of suspension to derive the total allowable home office overhead costs. The best known of these formulas is the Eichleay formula, because it is the judicially recognized formula for federal construction contracts.

### *Conclusion*

It is important when analyzing the contractor's requested compensation to focus on whether the costs that are claimed are truly limited to costs arising from performing the work in question. Costs not directly related to the circumstances of the change order should not be allowed.

## Elements of change order compensation

The two types of available damages, time and money, are calculated differently. Each is explained in more detail below.

Time-related damages consist of extensions to the contractually specified completion date or an appropriate number of days added to the contractually specified contract time. These damages are based on the analysis of the schedule and other relevant materials as discussed in an earlier section of this chapter which dealt with analyzing delay proposals.

The calculation of how much additional money to pay the contractor involves two steps. The first is adding up the allowable costs in the various categories described above. The second is comparing that total to an appropriate base. The base differs depending on the type of proposal. If the proposal is for extra work, the appropriate base is zero since the contractor would not have any costs associated with that work in the absence of an instruction to perform it. If the proposal is for lost productivity, the actual labor costs have to be compared to what the contractor's labor costs would have been if the condition causing the loss of productivity had not existed.

It is important to be alert for total cost proposals. A total cost proposal is one in which the contractor seeks reimbursement for all costs incurred. The contractor's position is, "My cost to complete this job was $1,000,000. After performing the work covered by the proposal, my costs to perform this job will be $1,100,000. Therefore, I'm entitled to $100,000 plus mark-ups."

This approach to determining the value of a change order proposal does not acknowledge any responsibility on the part of the contractor. For this reason, the total cost approach to valuing change order proposals and claims is viewed by courts with great skepticism. Proposals which have a total cost approach should be resisted. The contractor should be required to recalculate the value of the proposal to reflect the costs and/or the time for which the owner is responsible.

## How to analyze change order proposals

The overriding goal when analyzing a change order proposal is to make sure that the contractor is not paid any money that it is not entitled to receive. No matter how interesting, analyzing change order proposals is not an academic exercise. It is a "cold light of day" business activity and a vital part of the owner's cost control efforts for the project.

The purpose of this section is to describe a process that will assist in the effective and efficient analysis of change order proposals submitted by the contractor. There are numerous possible approaches to this process; the one suggested here has proven effective in analyzing a large number of change orders and claims, ranging in value from a few hundred dollars to 60 million dollars. The process involves seven steps as listed in Exhibit 9.21.

---

**Exhibit 9.21   Process for analyzing change order proposals**

1   Review the contractor's change order proposal;
2   analyze applicable contractual requirements;
3   evaluate the contractor's factual position;
4   determine entitlement;
5   calculate the appropriate compensation;
6   negotiate the appropriate value(s); and
7   prepare a formal, written response.

---

Each of these steps is described in greater detail below.

### *Review the contractor's change order proposal*

Upon receipt of the proposal, it is important to read it very carefully. Every fact alleged by the contractor and the contractor's theory underlying the proposal should be carefully scrutinized in order to prepare for the subsequent steps. The theory of the proposal is the basis on which the contractor believes it is entitled to recover.

### Analyze applicable contractual requirements

There are two types of contractual requirements: procedural and substantive. It is important to analyze both of these requirements. Each construction contract specifies the steps the contractor has to take to submit change order proposals. These are the procedural requirements, and they typically involve the following.

- *Submit notice to the owner* The contract will require that the contractor send the owner written notice of its intent to submit a change order proposal within a specified number of days following the event which the contractor believes gives rise to its proposal.
- *Submit change order proposal* The contract will require that within a specified number of days following the submission of the required notice, the contractor must submit the change order proposal.

The first step is to consider whether the contractor has complied with the procedural requirements of the contract. The review of compliance with the procedural requirements under provisions similar to those described above, involves determining the answers to the following questions.

- *Was the required notice submitted timely?* If the contractor has 30 days from the event to submit the notice of intent to submit a proposal, was it submitted within the 30-day window? It is important to note whether the contractually established period refers to working days or calendar days.
- *Was the change order proposal submitted on time?* The contractor has a specified number of days following the submission of its notice to submit the actual change order proposal. Was the proposal submitted within that time frame?

The provisions that set forth these various requirements also address the need for the proposal to be fully documented. However, whether the proposal is properly documented is not a procedural issue because proper documentation depends on what the contractor is claiming, which is a substantive issue.

The substantive contractual requirements relate to how the project will be constructed, and how the project will be managed. In general, the requirements relating to how the project will be built are contained in the plans and specifications. For example, a change order claiming the contractor incurred extra costs in connection with the installation of air ducts because the HVAC system was modified by the engineer would require review of the HVAC plans and specifications. These relate to how the project is to be built.

The requirements relating to how the project will be managed are often contained in the general conditions. For example, the owner may defend against a delay claim on the basis that the contractor's schedule does not comply with the contract's scheduling requirements. This would be a reference to provisions related to how the project will be managed.

When analyzing the contract, certain techniques are helpful. These include identifying specific issues; examining the contract thoroughly; reviewing key control provisions; and reading provisions in the context of the proposal. Each is discussed in turn.

In order to most effectively evaluate contractual provisions, it is important to identify the specific issues being considered. To identify the issues, start by reviewing the type of proposal the contractor submitted. The next step is to establish what the contractor must prove to support the proposal. This produces a first group of issues. Then there should be consideration of possible owner defenses. Determining which defenses may be applicable and what the owner has to demonstrate to prevail produces a second group of issues. For example, if the contractor claims that certain work is extra work, the issues will be (1) is the work really extra work; and (2) did the owner instruct the contractor to perform the work.

The owner may be intending to assert that the work in question was correcting defective work and therefore not extra work and not the owner's responsibility. The issues in connection with that line of defense are (1) what are the applicable contractual provisions; and (2) in what ways does the work in question not comply with those provisions.

In any given situation, there may be several contractual provisions that apply. It is important to identify all of them. In many contracts, provisions will contain more than one important point. At least some of these points may not be identified in the provision's title. Therefore, the contract must be read very thoroughly to find all applicable provisions.

Consideration should be given to the plans and specifications that relate to the work in question. Care should be given to check all relevant provisions. If the work in question relates to wiring the HVAC equipment, the plans and specifications related to both HVAC and electrical work should be reviewed.

There are certain contractual provisions that are intended to assist the owner to control the project. These provisions should be reviewed to determine if any of them are relevant to the proposal under consideration. Examples of such provisions are listed in Exhibit 9.22.

---

**Exhibit 9.22   Owner control provisions**

- The contractor is responsible for all the acts and omissions of its subcontractors and suppliers;
- the contractor must provide a complete, functional project;
- payment by the owner to the contractor does not relieve the contractor of its responsibility to correct defective work;
- the only remedy for delay is additional time;
- a waiver by the owner must be in writing to be effective and applies only to the matter covered by the writing, and not to any other similar or different matters; and
- an executed change order includes all the additional costs and/or time that the contractor will receive for the transaction covered by the change order.

---

The important question is not what a contractual provision may mean in theory; rather, the question is how the provision applies to the specific set of facts involved in the proposal under consideration.

For example, failure to properly specify the wiring of the HVAC equipment is too significant a problem for the owner to claim that the provision requiring a complete, functioning project applies. However, if the wiring is properly specified but the device connecting the wiring to a single piece of HVAC equipment is not fully shown, that provision may apply.

### Evaluate the contractor's factual position

Even if the contract supports entitlement if the factual basis of the proposal as stated by the contractor is accurate, it is important to evaluate whether the contractor's statements are in fact accurate. This involves confirming the work in question, utilizing construction inspection, utilizing the schedule, reviewing the contractor's documentation, and developing a conclusion.

The first step is to confirm in detail the work that is the basis for the contractor seeking extra money and/or time. The purpose of this step is to determine which work is, or should, be the subject of the proposal. This step is important because the contractor's description of the work may include work that is not appropriately part of the proposal.

In some cases the contractor will specifically identify the work. This is likely where the contractor asserts that the work in question is extra work. In such a case, the work that is claimed to merit additional money and/or time will be clearly described.

Where the contractor is claiming acceleration, the extra money sought relates to how certain work was performed, rather than to work clearly marked as extra. It is necessary to know which portion of the work is claimed by the contractor as having been accelerated.

An active construction inspection program will provide the information necessary to evaluate the contractor's claim relative to the work. The information will exist both in the form of documentation, and in the more informal but equally important form of the intimate knowledge of the project's progress by the person(s) performing the regular inspections.

- *Creating documentation* The person performing the inspection will be creating documentation based on their inspection activities. By keeping a diary, completing daily reports, drafting memos, and taking photographs/video, the person performing the inspection will develop documentation that can be used to analyze the contractor's proposal. This information is of great assistance in evaluating the contractor's description of the work and/or how it was impacted.
- *Knowledge from observation* Beyond the documentation, the person doing the inspection will develop an in-depth knowledge of the project's progress. The observations, impressions, and recollections of the person will also be useful in evaluating the contractor's position.

The baseline schedule and the updated schedules as submitted by the contractor are a critical source of information. They will show:

- when the work in question was supposed to be performed, either originally or as updated;
- when the work was actually performed; and
- how long the work took to perform.

If the baseline schedule and the periodic updates are submitted by the contractor and reviewed by or on behalf of the owner, then the information shown in the most recently approved schedule update should not be disputed by either party.

There are two standards that the contractor's documentation must meet. The first is the requirements set forth in the contract for documenting change order proposals. The second is relevance and strength of the documentation in terms of supporting factual assertions made by the contractor.

The contract will address the documentation that the contractor is required to submit to substantiate a proposal. Some contracts use general language to the effect that the contractor must submit sufficient documentation to substantiate the proposal. Others are more specific; for example, requiring material costs to be documented by the material supplier's invoices. The contract will require any proposal asserting a delay and seeking additional time to be accompanied by a schedule analysis. Whatever the contractual requirements are, the contractor's documentation must meet them.

The contractor's documentation must support its factual assertions. Some examples of commonly made assertions and the type of documentation that would support the assertions follow.

- *Labor costs* Certified payrolls will show hours worked and hourly wages; for labor burden, an itemized calculation showing how the percentage utilized was developed.
- *Material costs* Invoices submitted by the material supplier.
- *Equipment costs* For owned equipment, an explanation of the contractor's standard accounting practice for determining and charging appropriate rates; for rented equipment, a copy of the rental agreement.
- *Subcontractor costs* Requisitions for payment, including the attached documentation, submitted by the subcontractor to the contractor.
- *Differing site condition* Photographs of the condition claimed as differing.
- *Suspension of the work* A copy of the letter from the owner instructing the contractor to suspend work.
- *Acceleration* A copy of the letter from the owner directing the contractor to accelerate; a copy of progress or other meeting minutes confirming the owner's verbal directive to accelerate.
- *Agreement by owner to pay for extra work* A copy of the letter from the owner instructing the contractor to perform the work in question, characterizing the work as extra, and/or expressing a willingness to pay for it; a copy of progress or other meeting minutes confirming the owner's statement that the contractor is to perform certain work, that the work is extra, and/or that the owner agrees to pay for the work.
- *Delay* A CPM schedule analysis demonstrating the delay.
- *Acquiescence by the owner* A copy of a letter from the contractor containing an assertion which the owner would be expected to contradict, and a copy of the owner's response to that letter which does not address or contradict the contractor's assertion.

In cases where the contractor's claimed costs are in question notwithstanding the submission of invoices, rental agreements, subcontractor requisitions for payment, certified payrolls, or other documentation, the owner should request copies of cancelled checks or evidence of electronic payment representing payment of the questioned costs.

The contractor's documentation should be checked for mathematical accuracy. The amount of the proposal should not exceed the documented costs plus the contractually allowed mark-ups.

Using the various pieces of information described above, the owner will be able to determine to what extent it agrees with the contractor's factual position. The owner may agree entirely, partially, or not at all.

### Determine entitlement

The next step is to conclude whether the contractor's proposal has merit. That determination should be made by answering the following questions.

- Focusing on the applicable contractual provisions, is there entitlement under the contract if the facts supporting the proposal are accurately stated by the contractor? If the answer is positive, go to the next question. If the answer is negative, proceed to the last step (preparing a formal response).
- Are the facts substantially as described by the contractor? "Substantially" because minor discrepancies are to be expected in matters as complicated as many change order proposals. If the answer is positive, proceed to the next step (calculating compensation); if the answer is negative, proceed to the last step.

### Calculate the appropriate compensation

If the proposal has merit, the next step is calculating the appropriate compensation. Using the information contained in the documentation, calculate the extra costs and/or additional time to which the contractor is entitled. It is important for the owner to develop an independent estimate of the appropriate compensation. It can be used as the basis of negotiation with the contractor, or, it can be presented on a take it or leave it basis.

### Negotiate the appropriate value(s)

Frequently, when a change order proposal has merit, the owner and the contractor will have different opinions on the appropriate value of the change order. The owner should make a good faith effort to resolve this difference through negotiation. To successfully negotiate the value of a change order proposal, the owner needs to have a clear understanding of what the change order is worth and why. That understanding allows the owner to assess the reasonableness of the contractor's proposed value and to decide what concessions it can reasonably make.

The posture in negotiating a change order proposal will depend on several factors. The first is the extent to which the contract clearly dictates the amount of the change order. For example, if the contract has unit prices, and the scope of the change order involves unit price work, then the value of the change order should be easily determined. If the contractor argues that the total should exceed the value of the number of units times the contractual unit price(s), the owner should have no reason to show flexibility.

The second factor is the reasonableness of the values used in the change order proposal. If the contractor clearly documents the costs involved with the work that is the subject of the proposal, there may be less justification for proposing different, or at least significantly different, values. On the other hand, if the contractor's values are aggressive and/or not well documented, the owner will want to challenge the value of the proposal and offer what it considers a more reasonable value.

The third factor is the history of the project. If the contractor has made obvious efforts to make the project successful (completed as designed and on budget and on schedule) the owner may choose to be accommodating when dealing with a given change order proposal. On the other hand, when the contractor appears to be attempting to increase its profit or make up for a potential loss by submitting a stream of change order proposals, the owner may choose to challenge a proposal's value whenever there is room for a supportable challenge.

In the end, the owner's objective should be to agree to the compensation to which the contractor is contractually entitled; not more, not less.

### Prepare a formal, written response

The last step in analyzing a change order proposal is to prepare the formal response. If the proposal has merit and the value or values have been agreed upon, the formal response is the preparation of the change order. It is recommended that the owner prepare change orders because the precise wording on change orders can be very important. This is particularly true in connection with the provision that precludes the contractor from recovering costs from circumstances which are already the subject of a fully executed change order.

When the change order is rejected, the formal response will be a letter from the owner to the contractor explaining why the change order is rejected. It is important when conveying the owner's position to the contractor that it be detailed and explicitly based on applicable contractual provisions. The provisions should be cited or quoted.

## Claims

It is recommended that any construction contract require that any request for additional money and/or time first be submitted as a change order proposal. Exhibit 9.23 illustrates such a provision.

---

**Exhibit 9.23   Claim must first be change order**

11.2 *Claims Must be Change Order Proposals* Unless approved by Owner in writing, no Claim shall be submitted unless the act, omission, or circumstance giving rise to the Claim shall have first been submitted as a change order proposal and acted upon pursuant to the provisions of Article 9.

---

There are several reasons for this requirement. First, it forces the contractor to submit its request, or at least its notice of intent to submit a request, soon after the event or circumstance giving rise to the request. This is the point in time when the owner has the most options as to how it wishes to resolve the issue. It also provides the owner the ability to have timely budget control and cost reporting.

Second, requiring an issue to be addressed initially as a change order means it will be dealt with first by people at the project level. These are the people who are the most familiar with the details of the issue and, on most projects, the people who share an interest in building the project, as opposed to posturing over issues.

Third, by seeking to resolve most issues at the project level, senior management of the owner and the contractor can focus on seeking to resolve the major, difficult to resolve disputes. These focused, relatively infrequent negotiations should allow both sides to bring a positive commitment to those negotiations requiring their involvement.

Owners should anticipate that claims will be more carefully assembled than the related change order proposal. Typically, claims are more thoroughly explained and more comprehensively documented. There are at least two reasons for this. First, contractors don't usually assign someone to prepare change order proposals during construction. They are prepared by people with other responsibilities. Therefore, preparation of most change order proposals receives minimum attention from the contractor's staff.

The second reason claims are more carefully prepared is that contractors realize that resolving a claim is the last opportunity to settle the issue at the project level. If the claim is not resolved, the next step is contractually specified dispute resolution. Whether alternative dispute resolution or litigation is used, it can rapidly become expensive. So, on the one hand, the contractor's position will be more articulately presented; on the other hand, the contractor has an incentive to settle the issue.

The requirements the contractor must meet for each type of claim are the same as for the corresponding change order proposal. The techniques for evaluating claims are likewise the same.

# 10 Closing out the construction contract

## Introduction

Closing out the construction contract is a critical phase of the project. It is the final opportunity for the owner to ensure that it receives all the deliverables to which it is contractually entitled. Many owners will attest that this process can be difficult, contentious, and annoying. The closeout process becomes adversarial if the contractor fails to stay focused on the project till it is completed.

There are various reasons why this happens. The most common include the contractor believing that fully meeting all contractual obligations for closeout will cause it to lose money; the contractor focusing on its next project(s) or on other projects with challenges; and the owner imposing requirements for closing out the contract which go beyond what the construction contract provides.

When approaching closeout of the contract, it is important for the owner to remember that the contractor's work is not just the physical construction. It also includes all the services required by the construction contract. Exhibit 10.1 illustrates an appropriate contractual definition.

---

**Exhibit 10.1   Definition of work**

2.1 *The Work* Contractor shall perform all of the work required by the Contract Documents ("the Work"). The Work shall include all the construction and services required by the Contract Documents and all other labor, materials, equipment, management, coordination, and other services to be provided by Contractor to fulfill Contractor's obligations. It is understood that the Contract Documents provide for a complete Project, and Contractor shall perform all work necessary and reasonably inferable from the Contract Documents to provide such complete Project for the Contract Price.

---

If the contractor is required to submit certified payrolls to prove compliance with prevailing wage requirements, submission of the payrolls is part of the work. If the contractor is required to submit as-built drawings at the conclusion of construction, that is part of the work.

There are four important aspects to closing out a construction contract. These include administering the punch list, awarding substantial completion, determining and assembling the required documentation, and determining and releasing final payment. Each will be discussed in turn.

## Administering the punch list

The punch list is a list of all work that has not been completed as required by the contract documents. The work may be incomplete or it may be defective (i.e., complete but not built as required by the contract documents). The relevant question for including the item of work on the punch list is whether the work complies with the applicable requirements of the contract documents. If the work does not satisfy those requirements, it should be included on the punch list.

It is important to note that anyone's subjective judgment about the work is not relevant. The fact that the designer, or even the owner, does not like what the work looks like, does not matter. If the work meets the contractual requirements, it must be accepted and cannot be placed on the punch list.

There are three important issues in connection with the punch list. The first is determining whose responsibility it is to prepare the first draft of the punch list. The second is whether to require the contractor to monetize the punch list. The third is how to handle multiple iterations of the list.

It is recommended that the contractor be required to prepare the first draft of the punch list. There are two reasons for this. First, the contractor has the most in-depth knowledge of which portions of the work are incomplete or defective. Second, if the relationship with the contractor is contentious, the contractor's initial draft of the list represents work that the contractor concedes is incomplete or defective, so those items will not be the subject of debate.

Upon receipt of the contractor's draft punch list, it should be reviewed by the designer and by the owner. These reviews are for the purpose of determining whether the punch list submitted by the contractor is complete. If the owner has used a construction inspector, that person's detailed knowledge of the work and his/her resulting ability to accurately evaluate the contractor's draft punch list is another return on the inspection investment.

Upon approval of the punch list, the contractor is required to perform the punch list work. When the contractor believes the work is complete, it requests an inspection by the owner. The owner typically involves the designer in all punch list inspections.

Another important issue is whether the contractor should be required to monetize the punch list (i.e., assign a dollar value to each item). The principal reason for requiring it is to provide an agreed upon amount the owner should withhold to fund the completion of the work. The main argument against this requirement is that the value of many items on the list can each become a dispute. The more efficient approach, according to this argument, is to simply specify an amount to be withheld, such as 10 percent of the contract value. Either approach can be effective as long as the amount of money withheld to fund the punch list work is (1) sufficient to fully fund the remaining work, and (2) large enough in the context of the total project value so that the contractor needs the funds to realize at least a portion of its profit.

The completion of the punch list work will often require several editions of the list and several inspections. The question is who should pay for the time involved in reviewing the draft punch lists and inspecting the work. In order to incentivize the contractor to focus on completing the work, the construction contract should specify that the contractor will reimburse the owner for any costs associated with administering the third and subsequent lists.

## Awarding substantial completion

Awarding substantial completion is an important part of the project closeout process. That is because under many construction contracts when the contractor achieves substantial completion, a number of clocks start to tick. These may include such things as when retainage is due to be paid out, when final payment is due, and when warranty periods start. Each of these events includes risks to the owner and is discussed here in more detail.

Retainage is withheld for various reasons. One of them is to ensure that there are sufficient funds available to cover the costs of finishing the punch list and other activities required to close out the contract. Paying all retainage to the contractor will eliminate the assurance that the funds will be available. It also eliminates the leverage that withholding funds provides, which is an important consideration if the contractor is not focused on completing the project. On the other hand, withholding these funds longer than necessary raises fairness issues and will be a disincentive for some contractors to bid.

The timing of final payment is important. Some public jurisdictions have statutes mandating that final payment be made within a specified number of days following substantial completion. Some private contracts have similar provisions. The risk in these circumstances is that the owner will have to pay the contractor the full contract price even if the work is not really complete. These provisions typically allow the owner to withhold the value of correcting incomplete or defective work. Nonetheless, there is some residual risk that the owner will end up paying the contractor more than the value of the work performed (i.e., retaining less than the value of correcting the problematic work). This reduces the incentive for the contractor to keep focused on completing the project. At a minimum, this increases the chances of a dispute with the contractor over how much the owner should be withholding.

The typical construction contract provisions creating warranties specify that the warranty period starts at substantial completion. It is important to consider this issue in the context of the total project. For larger projects, the warranties for items that were installed early in the project may expire before the total project is available for use by the owner. The most effective way of dealing with this is to specify that warranties begin at substantial completion of the entire project, not at the substantial completion of the relevant subcontractor's work. Another way to deal with this issue is to specify that the warranties don't start till the owner has beneficial use of the building.

Because of these risks, it is very important for owners to control the award of substantial completion. The key to achieving this control is a favorable definition of substantial completion. Favorable means two things. First, the definition does not lead to the contractor being able to claim substantial completion before the owner can really use the facility in the manner the owner envisioned. Second, favorable means allowing the owner to keep control of the substantial completion process by specifying that substantial completion only takes effect when it is granted by the owner in writing.

Public jurisdictions may define substantial completion statutorily. For private projects, what constitutes substantial completion will be defined in the contract. Exhibit 10.2 gives the definition for substantial completion as shown in the contract in Appendix C.

This definition requires that before substantial completion can be awarded, the owner must be able to use the facility in the way the owner intended it to be used, and all government approvals required for occupancy have been obtained. At that point, the value of incomplete or defective work should be minimal and the start of warranty periods should coincide with the start of the owner's use of the facility.

With a provision such as this in the contract, the owner will want to ensure that all the conditions of the provision are satisfied before granting substantial completion. This can be more challenging than it appears. For example, some contractors will claim that the facility is available for its intended use if the lights go on, even if critical low voltage electrical

---

**Exhibit 10.2   Definition of substantial completion**

4.2 *Substantial Completion* The Work shall be deemed to be substantially complete
   when Contractor has substantially completed the Work and Owner determines
   that it is able to occupy and use the Project for its intended purpose(s)
   notwithstanding that minor amounts of the Work remain to be completed or
   corrected ("Substantial Completion"). Owner's ability to occupy and use the
   facility shall mean that all building systems are fully operational, that the facility
   is readily useable by Owner for its intended purpose(s), that Contractor has
   obtained a certificate of occupancy or, with owner's written approval, a temporary
   certificate of occupancy; and that any other requirements imposed by any
   authorities having jurisdiction have been satisfied. Notwithstanding any other
   provision of this Paragraph or of the Contract, Substantial Completion shall not
   occur until granted in writing by Owner.

---

systems are not fully installed or operational. To protect itself from the risks discussed above,
the owner should vigorously enforce the contractual requirements for substantial completion.

To further protect against these risks, the owner should not take any actions inconsistent
with the requirement that substantial completion only takes effect when granted by the
owner in writing. The owner should not acknowledge substantial completion in informal
conversations or in meetings. The owner's position should always be that when the contractor
has satisfied the conditions for substantial completion, it will promptly issue a written
statement to that effect, but not before then.

## Determining and assembling the required documentation

There are three tasks associated with assembling required documentation. These include
identifying what documentation is required to be submitted by the contractor; making sure
the contractor submits the required documentation; and determining that the submitted
documentation meets the relevant contractual requirements.

The first step, identifying the required documentation, should be relatively straight forward if the owner required the schedule of deliverables discussed in Chapter Three. The
schedule should include the documents required to be submitted at closeout. Exhibit 10.3
lists documents commonly required at closeout.

Each of the documents listed in Exhibit 10.3 is important to the owner. It is necessary
for the owner to review each document to ensure it satisfies all applicable contractual
requirements.

As-built design documents illustrate and describe how the project was actually built. This
information, including the location of studs, pipes, conduits, and other parts of a building not
readily observable, is very important for property management staff charged with keeping
the building operating. It will also be important if the building is to be modernized at some
point in the future. For those reasons the construction contract generally has requirements
related to completeness and accuracy of the as-built drawings to be submitted by the
contractor. If the owner has invested in construction inspection, this will be another return on
that investment because the inspector will have a very good sense if the as-built design
documents are accurate and complete.

**Exhibit 10.3   Commonly required documents at closeout**

The following documents are typically required at the closeout of a construction contract:

- as-built design documents;
- warranties;
- operations and training manuals;
- evidence that required training for equipment use has been conducted (including training video or disk if required);
- full set of material testing reports;
- commissioning final report;
- final lien waivers; and
- release of claims.

Warranties provide the owner with protection for a reasonable period of time against defective materials and equipment. They typically apply to roof materials, windows, exterior cladding (particularly if the cladding is a system attached in some way to the building structure), mechanical equipment, electrical equipment, elevators, and other building systems. Contractual provisions usually address the duration of the protection, to whom the protection runs (usually, the owner), and what the obligation is if the product does not act as specified. Warranties as submitted by the contractor must be carefully reviewed to ensure they comply with the contractual requirements.

Every construction contract will also contain a warranty related to workmanship. This applies to all physical work performed under the construction contract, including the installation of materials and equipment. This is usually a one-year warranty created by the contract. There are no separate submittals associated with this warranty.

Operations and maintenance (O&M) manuals explain how to operate and maintain the building equipment to which they apply. They are prepared by the equipment manufacturer and submitted by the contractor. The O&M manuals will be important to the property management staff because they explain how to operate and maintain the building systems. These manuals generally are for the mechanical, electrical, security, and other building systems.

Many construction contracts, particularly those for larger projects, go beyond O&M manuals. They require training for building operators. This requirement is designed to prevent damage to sophisticated, expensive equipment due to operator error. The requirement will often include illustrated training materials in the form of a video or devices for computer use, such as a disk or thumb drive. The required submission at closeout is evidence that the training was conducted as required and copies of the video, disk, or drive.

Commissioning is a key process in most building projects of any size. This includes new construction projects and projects involving the modernization of existing buildings. For those not familiar with the process, commissioning is the process of making sure the building systems are the most effective ones to meet the owner's program for the building, and that, following installation, the systems do actually operate as specified in the contract documents.

Commissioning is usually performed by one of two entities. On larger, more complicated projects, the owner will often use an independent commissioning agent contracted directly to the owner. The other approach is to make the mechanical subcontractor responsible for commissioning. This arrangement is more common on smaller, less complicated projects.

The submission at close out will be the final commissioning report. This report evidencing successful completion of the commissioning process is very important to the owner because it provides assurance that the building systems will operate as intended.

As part of the closeout process, the owner will want to know that all subcontractors have been paid the total amount that each subcontractor is entitled to under its subcontract. The owner will also want the contractor to agree that the only amount the owner still owes the contractor is the agreed upon value of the final payment. The most common document to be used to achieve this assurance is a lien waiver. This document says that in consideration for the agreed upon final payment, the contractor agrees it will not place a mechanics or any other type of lien on the project.

The construction contract usually also requires lien waivers from the subcontractors. These will say that in consideration for full payment by the contractor, which the subcontractor acknowledges having received, the subcontractor agrees it will not place a mechanics or any other type of lien on the project. Public projects cannot be liened. For that reason, public owners need a differently worded form, but the concept will be the same. The public agency acting as owner will want to be sure all subcontractors have been paid all monies owed to them, and that the contractor agrees it is owed nothing in addition to the agreed upon final payment.

The owner wants these assurances to lock in the final cost of the project. That is a necessary step in determining the amount of the permanent financing. For self financed projects, it is a necessary step for successful budget management.

There is another reason the owner should require these waivers. They represent a legally binding agreement on how much is owed to the contractor (the agreed upon final payment) and its subcontractors (nothing). These waivers are designed to preclude later attempts by the contractor to obtain additional sums because the project was not as profitable as originally planned or for any other reason.

The submission at closeout is the contractor's lien waiver and the lien waivers of each subcontractor. These documents must be executed by a legally authorized representative of the entity submitting the waiver.

The construction contract will often require a second document aimed at clearly defining the amount owed to the contractor. This document is a waiver of claims. This document says that in consideration for payment of the final payment in the agreed upon amount the contractor waives all claims against the owner. There is one caveat. This form typically says the contractor agrees to waive all claims "except those listed below." The contractor is required to list the one or more claims it intends to pursue beyond final payment, the proposed value of each claim, and the basis for each claim. This form should be written so that it covers all claims that have arisen on the project, and precludes later, separate claims by subcontractors. Even if the contractor asserts one or more claims, the claim waiver form is still important to the owner. That is because the contractor relinquishes all claims that are not listed on the waiver of claim form.

The submission at closeout is the waiver of claim form. It must be executed by a legally authorized representative of the contractor.

## Determining and releasing final payment

The last step in closing out the construction contract is determining the appropriate amount of the final payment and delivering the payment in that amount to the contractor.

Calculation of the amount starts with establishing the final contract price. This is the original contract price as modified by all change orders on the project agreed to by the owner. In other words, it is the original contract price plus the aggregate value of all positive change orders minus the aggregate value of all deductive change orders. This net number is the total contract price.

A word of caution is in order in connection with the calculation of the total contract price. Contractors famously dislike credit (i.e., negative) change orders. They can be very slow to submit the final proposed version of credit change orders. The owner should not allow itself to be pressured into establishing the final contract price till all credit change orders have been fully negotiated and executed.

The maximum amount of the final payment—without regard to outstanding claims, if any—is the total contract price minus the aggregate value of all payments made to the contractor to date. This is the maximum value because various deductions from that amount may be contractually justified. A common example is the value of incomplete or defective work that, following the punch list process, the owner has elected to finish using its own forces or a separate contractor. Another common example is the monetized value of deliverables owed by the contractor but not submitted, or the submissions are incomplete (e.g., as-built drawings and the as-built schedule). A third example involves the monetized value of a determination that an approved substitute material or equipment does not, in fact, perform substantially the same as the originally specified item.

On some projects, even these items may be handled by deductive change orders. On others, they will be subject of a negotiation with the contractor over the value of the final payment. The latter approach is acceptable as long as the elements of the negotiation and the final resolution are carefully and comprehensively documented.

# Appendix A

## Agreement for design and construction administration services

This is an Agreement ("Agreement"), made this _____ day of _____, 20__, for valuable consideration by and between _____ ("Owner"), having a principal place of business at _____, and _____ ("Designer"), having a principal place of business at _____, for the performance of professional design and construction administration services by Designer.

    For good and sufficient mutual consideration the parties agree to the terms and conditions as set forth in the Agreement.

### Article 1 The Project

1.1    The project which is the subject of this Agreement is _____ ("Project").

1.2    The Project is located at _____ and consists of _____ ("Site").

### Article 2 Designer's Responsibilities

2.1    *Designer's Services* Designer's services shall consist of all services required by this Agreement, whether performed by Designer, its subconsultants, or any other person or entity performing services on behalf of Designer. Designer shall be responsible for all the acts and omissions of its subconsultants and any other person or entity performing services on behalf of Designer

2.2    *Designer's Personnel* Designer shall provide a sufficient number of personnel with all the necessary design and construction administration professional skills and disciplines, and the necessary project management skills and experience to perform Designer's obligations under this Agreement.

2.3    *Design Schedule* Designer shall perform its services in accordance with the Schedule for Design Services ("Design Schedule") included in this Agreement as Exhibit A. Designer acknowledges that time is of the essence in the performance of this Agreement. Designer further acknowledges that it has reviewed the Design Schedule and that it is a reasonable schedule. The Design Schedule shall only be extended for causes that are beyond Designer's control.

2.4    *Notice to Proceed* Designer shall commence the performance of its services pursuant to this Agreement only upon receipt of a written Notice to Proceed issued by Owner. If such Notice is a partial Notice, Designer shall only perform those services covered by such partial Notice.

2.5   *Standard of Care* Designer shall perform the Scope of Services set forth in this Agreement using the highest level of professional care applicable to the Project. Such highest level of professional care shall be utilized by the Designer in the performance of each part of its Scope of Services as set forth in Article 3.

2.6   *Designer's Proposal* The proposal submitted by Designer ("Designer's Proposal") is included in this Agreement as Exhibit B. If not included in Designer's Proposal, Designer shall submit a copy of its internal Quality Assurance Program, which shall be incorporated into this Agreement as part of Exhibit B. All representations made in the Designer's Proposal are included in this Agreement, provided that to the extent that any provision of the Designer's Proposal is inconsistent with the provisions of this Agreement, the provisions of this Agreement shall prevail.

2.7   *Named Individuals* Designer's Proposal describes the participation and responsibilities of certain named individuals (individually and collectively, "Named Individuals") and Designer acknowledges that the participation of the Named Individuals as described in Designer's Proposal was a material inducement for Owner entering this Agreement. Therefore, Designer shall not replace any of the Named Individuals without Owner's approval. The failure of any Named Individual to perform the responsibilities described in the Designer's Proposal, for reasons within the control of Designer, shall be considered a material breach of this Agreement.

2.8   *Designer's Project Manager* The person named in Designer's Proposal as the Project Manager shall have full authority to carry out Designer's responsibilities as specified in this Agreement.

2.9   *Owner's Program* By submitting its Proposal, Designer represents that it has fully familiarized itself with the Owner's program, its objectives for this Project, and the Owner's Budget described in Paragraph 5.4. Failure to so familiarize itself with such program, objectives, and Owner's Budget, will not reduce Designer's responsibility for meeting such program and objectives, and for preparing plans and specifications which allow the Project to be built for an amount no greater than that set forth in Owner's Budget.

## Article 3 Scope of Services

3.1   *Design Documents* Designer shall prepare all plans and specifications (collectively, "Design Documents") necessary for the construction of the Project. Such Design Documents shall be complete, in sufficient detail to allow the contractor selected by Owner ("Contractor") to construct a complete Project and to construct the project in full compliance with all applicable federal, state, and local laws, regulations, ordinances, and building codes.

3.2   *All Necessary Services* Designer's services shall include all architectural, engineering, construction administration, and other services necessary to perform the Scope of Services.

3.3   *Coordination* Designer shall take all necessary steps to ensure that the preparation, completion, and administration during construction of the Design Documents are fully coordinated. Coordination for the purposes of this Paragraph shall mean the timely and appropriate involvement of each of the various design disciplines throughout the design and construction of the Project.

3.4   *Permits and Approvals* Designer shall assist Owner to obtain all government agency approvals and permits necessary to proceed with the Project. Such assistance shall include preparing applications, drawings, and other required materials; attending community meetings and hearings of public bodies with jurisdiction over the Project; and performing any other activities reasonably required to obtain such permits and approvals. Any of these activities required of Designer for the third and subsequent attempts to obtain the same permit, provided the rejection of the first two applications were not the responsibility of Designer, shall constitute Additional Services for the purposes of Paragraph 7.2.

3.5    *Schematic Design Documents* Designer shall provide Schematic Design Documents that establish the conceptual design of the Project illustrating the scale and relationship of the Project components. The Schematic Design Documents shall include a conceptual site plan and preliminary building plans, sections, and elevations. Preliminary selections of major building systems and construction materials shall be noted on the drawings or described in writing. Owner shall approve the Schematic Design Documents in writing.

3.6    *Design Development Documents* Following the approval of the Schematic Design Documents, Designer shall commence work on the Design Development Documents. The Design Development Documents shall be based upon, and shall constitute refinement of, the Schematic Design Documents. The Design Development Documents shall establish the scope, size, and appearance of the Project, as well as the make-up, shape, and relationships of its major components, utilizing plans, sections and elevations, typical construction details, and equipment layouts. The Design Development Documents shall include specifications that identify major materials and systems and establish in general their quality levels. Owner shall approve the Design Development Documents in writing.

3.7    *Construction Documents* Following the approval of the Design Development Documents, Designer shall commence work on the Construction Documents. The Construction Documents shall set forth in detail the requirements for construction of the Project. The Construction Documents shall include Drawings and Specifications that establish in detail the quality levels of materials and systems required for the Project. Designer shall prepare bidding or proposal forms and the Construction Contract, including the plans and specifications and the Special and/or Supplementary Conditions, if any, which shall be included in the package provided to bidders. Owner shall approve the Construction Documents in writing.

3.8    *Cost Estimates* Based on the Owner's Budget provided pursuant to Paragraph 5.4, Designer shall provide cost estimates to Owner as part of the submission of Schematic Design Documents, Design Development Documents, and Construction Documents. Each such estimate shall be broken down by trade. Each such cost estimate shall be approved by Owner as part of its approval of the Schematic, Design Development, and Construction Documents.

3.9    *Exceeding Cost Estimates* If Designer has reason to believe that the cost estimate to be submitted with the next set of Design Documents is at least five percent (5%) greater than the amount approved by Owner with the previous set of Design Documents, Designer shall inform Owner of such projected increase of cost immediately upon concluding such increase is likely. Owner and Designer shall consider appropriate alternatives, including, but not limited to, increasing the Owner's Budget, modifying the design, and/or changing materials and equipment. Any modifications to the Design Documents made because of such projected increase in costs shall be performed at no additional cost to Owner unless such modifications result solely from changes to the Design Documents directed by Owner for reasons other than meeting Owner's Budget.

3.10 *Owner Approval for Increasing Budget* In no event shall the Owner's Budget be increased unless approved in writing by Owner. If Designer believes it is necessary to exceed the Owner's Budget, it shall submit to Owner a written request for approval which shall include an explanation of the reasons for the request, a description of Designer's efforts to avoid the request, and the new proposed amount for the Owner's Budget. Owner shall respond to such request in writing by either rejecting such request or by approving it at the requested amount or at some alternative amount determined by Owner.

3.11 *Bidding and Procurement* Following the development of the Construction Documents, Designer shall prepare and, if instructed by Owner, place advertisements that include bidding and procurement information, including the scope of the Project; the time, place, and conditions of the bidding; and other appropriate information.

3.12 *Evaluating Bids* Designer shall evaluate the bids submitted for the construction of the Project and make recommendations to Owner on selecting the contractor. Such assistance shall include, but not be limited to, calling references or otherwise obtaining information on bidders' past performances; analyzing bids to determine if they are insufficient, exorbitant, unbalanced, or otherwise objectionable; reviewing information concerning such bidders available from public agencies; and writing a report describing the information obtained through such activities.

3.13 *Lowest Construction Bid Exceeds Budget* If the value of the lowest construction bid exceeds the Owner's Budget, Owner at its sole discretion, shall elect to increase the Owner's Budget, rebid the Project, terminate the Project, or order the redesign of the Project to meet the Owner's Budget. If Owner instructs Designer to redesign the Project because the lowest bid exceeded the Owner's Budget, such redesign shall be accomplished at no additional charge to Owner. Owner shall participate in such redesign to the extent necessary to ensure that such redesign meets Owner's objectives for the Project.

3.14 *Construction Administration* Following bid award, Designer shall administer the contract entered into between Owner and Contractor ("Construction Contract"). Designer acknowledges the importance of its effective performance of its construction administration responsibilities to the successful completion of the Project, and further acknowledges that Designer's representations with respect to its capabilities in construction administration were a material consideration in Owner's entering into this Agreement. Designer's construction administration responsibilities are described in Paragraphs 3.14 through 3.24. To the extent such Paragraphs do not explicitly state what effective performance means, these Paragraphs shall be deemed to include the requirements that Designer perform the responsibilities of such Paragraphs expeditiously, with attention to detail, with detailed knowledge of the nature and extent of Contractor's performance, and with emphasis on enforcing the requirements of the Construction Contract.

3.15 *Progress Meetings* Designer shall convene and preside at weekly progress meetings to be held at the site of the Work. Such meetings shall be attended by appropriate personnel of Owner, Designer, and Contractor. Designer shall take the minutes of each progress meeting and distribute the minutes to all attendees. Designer's minutes, to be taken in a format approved by Owner, shall be the official record of each such meeting and be issued to attending parties within three (3) days of the meeting date.

3.16 *Requests for Information* Designer shall consider and respond to any requests for information submitted by Contractor. Such responses shall be in sufficient detail for the Contractor to understand and implement Designer's instructions and in sufficient time so as not to delay the progress of the Work.

3.17 *Review of Submittals* Designer shall review all shop drawings, product samples, and other submittals by Contractor required by the Contract Documents. Such review shall be in sufficient detail to confirm that such shop drawings, product samples, and other submittals comply with all applicable provisions of the Contract Documents. Designer shall mark each such shop drawing, product sample, or other submittal as either "Approved," "Approved as noted," "Resubmit with changes as indicated," or "Rejected." Designer may substitute its standard wording for those in the previous sentence, if approved in writing by Owner, provided that Designer shall agree that whatever words are substituted for "Approved" shall have the legal effect of approval.

3.18 *Review of Contractor's Schedule* Designer shall review and recommend approval or disapproval to Owner of the baseline schedule and schedule updates required by the Construction Contract. Such approval shall be limited to determining that such schedule submissions provide a realistic approach to completing the Project within the Contract Time, as defined in the Construction Contract, and that the updates also accurately reflect the progress of the Work.

3.19 *Inspection of the Work* Designer shall regularly inspect the progress of the Work. Such inspections shall be daily, unless some other interval is approved in writing by Owner, and shall be of sufficient duration to provide Designer with detailed knowledge of the progress of the Work. Immediately following each such inspection, Designer shall complete a field report in a form approved by Owner. Copies of such field reports shall be attached to Designer's Monthly Report described in Paragraph 3.25, except that a copy of any field report which describes an actual or potential critical problem shall be provided to Owner no later than the end of the day on which the inspection was conducted.

3.20 *Defective Work* Designer shall, as part of its inspections of the Work, identify defective work and shall notify Owner in writing of such defective work. Copies of such written notifications shall be provided to Contractor. Defective work for the purposes of this Paragraph shall mean work that does not comply with the requirements of the Contract Documents.

3.21 *Schedule of Values* Designer shall review the initial Schedule of Values submitted by Contractor pursuant to the Construction Contract. Such review shall consider if the listed activities and the dollar values assigned to each such activity are reasonable. Following such review, Designer shall recommend to Owner whether to approve such Schedule of Values or to require modifications thereto.

3.22 *Review of Requisitions for Payment* Designer shall review each Requisition for Payment ("Requisition") submitted by Contractor pursuant to the Construction Contract and shall recommend to Owner what action to take with respect to each such Requisition. Such review shall consider the mathematical accuracy of the Requisition; the nature and extent of the supporting documentation; the values and percentages shown in the Schedule of Values required to be submitted with the Requisition by the Construction Contract; the amount of the Work performed and the extent to which it complies with the requirements of the Construction Contract; and the updated project schedule reflecting the progress of the Work to date.

3.23 *Evaluation of Change Orders* The Designer shall review all change order proposals submitted by Contractor. Such review shall focus on the extent, if any, of Contractor's contractual entitlement to such change order under the Construction Contract, the adequacy of Contractor's documentation, the status of the work that is the subject of the change order proposal, and any other matters relevant to evaluating the change order proposal. Following such review, Designer shall recommend in writing to Owner what action to take with respect to such change order proposal and for what reasons.

3.24 *Evaluation of Claims* The Designer shall analyze all claims submitted by Contractor. Such review shall focus on the extent, if any, of Contractor's contractual entitlement to such claim under the Construction Contract, the adequacy of Contractor's documentation, the status of the work that is the subject of the claim, and any other matters relevant to evaluating the claim. Following such analysis, Designer shall recommend in writing to Owner what action to take with respect to such claim and for what reasons.

3.25 *Monthly Reports* Designer shall submit to Owner by the fifth (5th) day of each month a monthly progress report ("Monthly Report") which shall describe the services performed during the preceding month, important issues to be resolved, Designer's progress compared to the Schedule contained in Exhibit A, amounts paid to Designer compared to the Agreement Sum, and what Designer expects to accomplish during the next thirty (30) days. Attached to the Monthly Report shall be the field reports required by Paragraph 3.19 for the preceding month.

## Article 4 Ownership of Work Product

4.1 *Instruments of Service* Designer shall own the plans, specifications, and other documents produced by Designer (collectively, "Instruments of Service") in the course of performing its services pursuant to this Agreement. Designer shall grant to Owner an exclusive license to use the Instruments of Service in connection with the design, construction, expansion, maintenance, and/or reconstruction of the Project. Owner's license shall not extend to, and Owner shall not use, the Instruments of Service for any other purpose.

4.2 *Limitation on Use* Designer shall not use the Instruments of Service for any purpose other than to provide its services pursuant to this Agreement without the advance written approval of Owner.

## Article 5 Owner's Responsibilities

5.1 *Program* Owner shall provide designer with a program or other statement of Owner's objectives ("Program") for the Project. The Program shall contain sufficient detail that Designer shall have a full understanding of Owner's expectations for the Project.

5.2 *Existing Documents* Owner shall provide Designer with all existing design documents, survey plans, geotechnical reports, and any other documents in Owner's possession which describe the current and/or past physical characteristics of, the legal boundaries of, and any other relevant information about the Project.

5.3 *Owner's Project Manager* Owner shall appoint a Project Manager who shall be authorized to carry out Owner's responsibilities under this Agreement.

5.4 *Budget* Owner shall develop a budget, which shall be known as Owner's Budget. The Owner's Budget shall represent the maximum amount available for the construction of the Project. The Owner's Budget shall increase only if such increase is approved in writing by Owner.

## Article 6 Payment

6.1 *Agreement Sum* Owner shall pay the sum of $_____ ("Agreement Sum") to Designer. The Agreement Sum shall be paid to Designer as provided in Paragraph 6.2.

6.2  *Payment by Monthly Requisition* The Agreement Sum shall be paid to Designer on a monthly basis with the value of the invoice calculated by multiplying the number of hours worked by each person performing the Scope of Services during that month by the designated hourly rate of each such person as set forth in Exhibit C. In no event shall such rates exceed those set forth in Exhibit C except pursuant to an Amendment as authorized by Article 7.

6.3  *Invoices* Designer shall submit invoices on or before the fifth (5th) day of the month following the month during which the services that are the subject of the invoice were performed. Owner may request any documentation reasonably necessary to determine if Designer is entitled to the amount set forth in Designer's invoice. Owner shall make payment to Designer within thirty (30) days of the receipt of an acceptable invoice from Designer. Such payment shall be for the full amount of the invoice, unless the amount is disputed by Owner, in which case, Owner shall pay the undisputed amount within such thirty (30) days.

6.4  *Reimbursable Expenses* The Agreement Sum shall include all expenses of Designer associated with the performance of its services pursuant to this Agreement. Any Amendment which increases the Agreement Sum shall likewise include all expenses associated with the performance of the services which is the subject of the Amendment. Designer shall not seek reimbursement for any expenses associated with performing its services. If Designer will incur expenses not reasonably foreseeable when this Agreement or an Amendment was executed, and such expenses will be incurred solely because of instructions given to Designer by Owner, Designer may seek advance written approval from Owner for reimbursement of such expenses, and if such approval is granted in advance by Owner, Designer may obtain reimbursement from Owner for such expenses.

## Article 7 Changes to the Agreement

7.1  *Amendments* Subject to Paragraph 7.3, the Agreement Sum, and any other terms of this Agreement, shall be modified only by written amendment ("Amendment") signed by both Owner and Designer. No modification to any term of this Agreement shall take effect until the Amendment containing such modification has been fully executed.

7.2  *Additional Services* Increases or decreases to Designer's Scope of Services shall be made only by Amendment to this Agreement. Services requested of Designer by Owner which are not set forth in Article 2 above shall be considered Additional Services for which Designer shall be entitled to an increase in the Agreement Sum. No increase in the Agreement Sum shall occur as a result of the performance of Additional Services unless Owner has given advance written approval for such Additional Services and such increase in the Agreement Sum.

7.3  *Unilateral Amendments* Notwithstanding any other provision of the Agreement, Owner may issue a unilateral (i.e., signed only by Owner) Amendment to Designer modifying the Scope of Services and/or any other provision of the Agreement, and Designer shall forthwith perform such Amendment. Such unilateral Amendment shall become binding as written on Designer and Owner unless Designer disputes such unilateral Amendment pursuant to the provisions of Article 11.

7.4  *Not Eligible as Additional Service* Notwithstanding any other provision of this Agreement, the following shall not constitute Additional Services: preparation of analyses of proposed change orders as required by Paragraph 3.23 until the value of such proposed change orders exceed twenty-five percent (25%) of the original Contract Sum; preparation of analyses of claims as required by Paragraph 3.24; and preparation and administration of the punch list. No service performed by Designer shall be an Additional Service if it is performed because of Designer's negligence, error, or failure to perform services in accordance with the requirements of this Agreement.

## Article 8 Insurance

8.1   *Required Policies* Designer shall at all times while performing services on this Project carry insurance policies at least sufficient to meet the following requirements with insurance firms authorized to provide such policies in the state, or country, in which the Project is located: a) General Liability Insurance: $1,000,000 per occurrence and $2,000,000 annual aggregate; b) Automobile Liability Insurance: $1,000,000 per person and $1,000,000 per accident; c) Workers' Compensation Insurance as required by statute; d) Employer's Liability Insurance: $500,000 per occurrence; and e) Professional Liability Insurance: $1,000,000 per claim and $1,000,000 annual aggregate. The policies required by clauses a–d shall name Owner as an additional insured.

8.2   *Certificates of Insurance* Designer shall, prior to commencing the performance of services on this Project and thereafter as required by Owner, submit to Owner certificates of insurance which substantiate that Designer has insurance policies in force which meet the requirements of Paragraph 8.1. Owner may require that any policy required by Paragraph 8.1 be submitted for its review.

8.3   *Cancellation* Designer shall notify Owner immediately if any policy required by this Agreement is cancelled for any reason, or if such policy is modified in any way the effect of which is that Designer does not have insurance in force as required by Paragraph 8.1. Following the completion of its services on this Project, Designer shall not cancel any policy required by Paragraph 8.1 unless it has given Owner thirty (30) days advance notice in writing.

## Article 9 Indemnification

9.1   *Personal Injury or Property Damage* To the fullest extent allowed by law, Designer shall hold Owner harmless and shall reimburse all costs incurred by Owner, including but not limited to the payment of damages, attorneys' and other professional services fees, and other expenses, as a result of personal injury or damage to personal property arising from the negligence, error, acts, or omissions of Designer or any persons or entities for whom Designer is responsible under the Agreement.

9.2   *Failure to Perform According to Agreement* To the fullest extent allowed by law, Designer shall hold Owner harmless and shall reimburse all costs incurred by Owner, including but not limited to the payment of damages, attorneys' and other professional services fees, and other expenses resulting from the negligence, error, acts, omissions, or failure of Designer or any persons or entities for whom Designer is responsible under the Agreement to perform its services in accordance with the requirements of the Agreement.

## Article 10 Termination

10.1   *Designer's Termination* Designer may terminate this Agreement for Owner's failure to make payment in the manner required by Paragraph 6.2 above by providing fifteen

(15) days advance written notice to Owner. If Owner fails to make payment or provide a reason why it has not made payment that complies with the requirements of this Agreement within such fifteen (15) days, Designer may terminate this Agreement at the end of such fifteen (15) day period. Designer may terminate this Agreement for any other failure by Owner to comply with the requirements of this Agreement by providing thirty (30) days advance written notice to Owner setting forth Owner's repeated or material failure to comply with the material requirements of this Agreement. If Owner fails to correct such failure or to provide a reason for such failure(s) that complies with the requirements of this Agreement, Designer may terminate this Agreement at the end of the thirty (30) day period.

10.2  *Owner's Termination* Owner may terminate this Agreement for Designer's repeated or material failure to comply with the material requirements of this Agreement by giving Designer seven (7) days advance written notice. If Designer does not correct such failure within such period, Owner may terminate the Agreement at the end of such seven (7) day period. Owner may terminate this Agreement for its convenience by giving Designer fifteen (15) days advance written notice.

10.3  *Payment to Designer* Upon termination, regardless of cause, Designer shall submit within ten (10) days of the effective date of the termination an invoice for the value of the work performed up to the effective date of the termination. If Designer was not terminated for any reason for which it bears responsibility, Designer shall be permitted to seek reimbursement for any reasonable costs solely attributable to such termination. Such costs shall not include anticipated revenues, lost profits, or any other anticipatory or consequential damages of any kind.

10.4  *Indemnification* If Designer is terminated by Owner other than for Owner's convenience, Designer shall hold Owner harmless for all damages and costs arising from Designer's termination, including but not limited to reprocurement costs, loss of use and/or revenue because of delayed completion of the project, attorneys' and other professional services fees, and other costs.

## Article 11 Disputes

11.1  *Resolution of Disputes* Disputes arising under this Agreement shall be resolved according to the provisions of this Article 11.

11.2  *Claim Notice and Statement* If Designer is aggrieved by an act or omission of Owner, it shall send Owner a written Notice ("Notice") stating the basis of its claim within twenty-one (21) days after the act or omission giving rise to the complaint. Designer shall submit a Statement of Claim ("Statement") within thirty (30) days after it submits its Notice. The Statement shall contain a detailed description of the claim and shall include all documentation necessary to clearly substantiate Designer's claim. Designer is solely responsible for documenting its claims. Designer acknowledges that the requirements of this Paragraph 11.2 are material provisions of the Agreement, and claims that are not pursued according to such requirements are waived.

11.3  *Owner's Response* Owner shall respond to Designer's claim in writing within thirty (30) days of receipt of Designer's Statement. If Owner cannot decide the claim within thirty (30) days, it shall send a written notice to Designer stating a date certain by when the claim will be decided.

11.4  *Litigation* If Designer is not satisfied with the Owner's response to a Statement, it may commence a legal action in a court of competent jurisdiction in the State of

_____. Compliance with the provisions of Paragraphs 11.2 and 11.3 shall be a precondition to the commencement of such action. Claims properly submitted pursuant to Paragraph 11.2 to which Owner has not responded at the time of final payment may be litigated at that time notwithstanding the absence of Owner's response pursuant to Paragraph 11.3.

11.5 *Alternative Dispute Resolution* Owner and Designer may agree to submit any dispute not resolved according to the provisions of Paragraphs 11.2 and 11.3 to mediation and/or arbitration. The rules for such mediation and/or arbitration shall be the current rules for construction mediation and arbitration of the American Arbitration Association.

## Article 12 Miscellaneous Provisions

12.1 *Integration* This Agreement shall represent the entire agreement between Owner and Designer and shall supersede any prior agreements, representations, and understandings, written or oral.

12.2 *Calendar Days* References to days in this Agreement shall mean calendar days unless otherwise stated. When a specified period would end on a Saturday, Sunday, or holiday, it will be deemed to end on the next business day.

12.3 *Law Governing This Contract* The provisions of this Agreement shall be construed according to the laws of the State of _____.

12.4 *Headings* The headings contained in this Agreement are for convenience only; they are not part of the Agreement and shall not define or limit the scope of any provision of the Agreement.

12.5 *Provision Unenforceable* If any provision of this Agreement is held to be unenforceable by a court of competent jurisdiction, the remainder of the Agreement shall remain in effect as if the invalid term had not been initially included.

12.6 *Conflict of Interest* Designer, by executing this Agreement, certifies that it has no current or future obligation to perform services that would constitute a conflict of interest with its obligations to Owner under this Agreement.

12.7 *Only Owner Liable* Designer shall seek to enforce the provisions of this Agreement only against Owner. No official, board member, employee (paid or volunteer), or agent of Owner shall incur any liability to Designer as a result of this Agreement.

12.8 *Notices* Written notices required by this Agreement to be sent by one party to the other shall be sent to the addresses first appearing in this Agreement to the attention of the receiving party's Project Manager. Either party may change the contractual notice address by providing written notice to the other party stating the new notice address.

12.9 *Records* Designer shall keep all plans, specifications, drawings, memoranda, correspondence, invoices, reports, and all other documents generated as a result of the services provided under this Agreement for a period of six (6) years following the date of final completion of the construction of the Project. Owner may request and shall receive copies of any such documents at any time during such six (6) year period.

Wherefore the parties make this Agreement under Seal as of the date first set out above.

| Owner: | | Designer: | |
|---|---|---|---|
| By: | _____ | By: | _____ |
| Title: | _____ | Title: | _____ |

# Appendix B
## Agreement for consulting services

This agreement ("Agreement") is entered into as of the ___ day of _____ , 20__ , for valuable consideration by and between _____ ("Owner"), having a principal place of business at _____ , and _____ ("Consultant"), having a principal place of business at _____ , for the performance of professional services by Consultant in connection with the design and construction of the _____ Project ("Project).

### Article 1 The Agreement

1.1 *Agreement* The Agreement shall consist of the terms and conditions set forth herein and the attachments listed in Paragraph 1.2 below, which attachments are all as fully a part of the Agreement as if repeated within the body of the Agreement.

1.2 *Attachments* The attachments to this Agreement shall include Attachments 1 through _____ as set forth below. Certain Attachments may be marked "Not Included," in which case such Attachments shall not be part of the Agreement.

| Attachment No. | Description |
|---|---|
| [etc., if needed] | |

1.3 *Priorities* In case of ambiguities, inconsistencies, or conflicts between any parts of the Agreement, the Agreement shall be interpreted on the basis of the following priorities:

a. written amendments to this Agreement as authorized by Article 6 of the Agreement;
b. the provisions of the Agreement; and
c. the provisions of the Attachments.

In the case of ambiguities, inconsistencies, or conflicts within the same document (i.e., within a, b, or c above), such ambiguities, inconsistencies, or conflicts shall be resolved in favor of the greater level of effort or contractual requirement.

1.4 *Entire Agreement* The Agreement shall represent the entire agreement between the parties. It shall supersede any oral agreements or representations by either or both parties, and it shall supersede any prior written agreements or representations by either or both parties.

1.5 *Term of Agreement* The term of the Agreement ("Term") shall be for the period starting on the date on which Consultant first performed its services on this Project and ending on the date the Project is finally complete as established in writing by Owner. The Term may be adjusted in accordance with the provisions of Article 6 or Article 9.

## Article 2 Owner's Responsibilities

2.1   *Program* Owner has provided Consultant with a program or other statement of Owner's objectives ("Program") for the Project. Consultant acknowledges that the Program contains sufficient detail that Consultant has a full understanding of Owner's expectations for the Project.

2.2   *Existing Documents* Owner has provided Consultant with all existing, and shall provide all future, documents in Owner's possession which are reasonably necessary for Consultant to perform its Scope of Services.

2.3   *Owner's Project Manager* Owner shall appoint a Project Manager who shall be authorized to carry out Owner's responsibilities under this Agreement.

2.4   *Procedures* Owner may elect to promulgate procedures for the management of the Project. Consultant shall follow such procedures as they may be applicable to Consultant's provision of services, requests for payment, or other aspects of its involvement on the Project.

2.5   *Review of Consultant's Work* Consultant's Deliverables (as defined in Paragraph 3.1) shall be subject to review, at such times and in such manner as Owner shall determine, by Owner, designers, other consultants retained by Owner, and contractors. Consultant agrees to work with Owner and such other persons or entities as may be designated by Owner in order to assure that the Project is designed and built in the manner most likely to achieve Owner's objectives for the Project.

## Article 3 Consultant's Responsibilities

3.1   *Consultant's Services* Consultant shall provide all services required by the Agreement ("Scope of Services"), whether performed by Consultant, its subconsultants, or any other person or entity performing services on behalf of Consultant. Consultant shall be responsible for all the acts and omissions of its subconsultants and any other persons or entities performing services on behalf of Consultant. Consultant's Scope of Services is set forth in *Attachment No. __*. Consultant shall perform and/or submit all tests, plans, reports, and other activities and documents (collectively, "Consultant's Deliverables") required by the Scope of Services in form and content reasonably satisfactory to Owner.

3.2   *Compliance with Applicable Laws* Consultant shall perform the Scope of Services in compliance with, and shall ensure that all of Consultant's Deliverables are in compliance with, all applicable federal, state, and local statutes, ordinances, codes, rules, and regulations.

3.3   *Consultant's Personnel* Consultant shall provide a sufficient number of personnel with all the necessary professional skills and disciplines, and the necessary project management skills and experience to perform Consultant's obligations under this Agreement. The participation on the Project of certain specific personnel ("Key Personnel") is of the essence of this Agreement. The Key Personnel shall consist of the Project Manager named pursuant to Paragraph 3.6 and _____. Consultant shall not add to or remove any of the Key Personnel from the team working on this Project without Owner's written consent.

3.4   *Schedule of Performance* Consultant shall perform its services in accordance with the Schedule of Performance ("Schedule") set forth in *Attachment No. __*. Consultant acknowledges that time is of the essence in the performance of the Agreement and that meeting the Schedule, including but not limited to completing Consultant's Deliverables as required by the Schedule, is vitally important to the Owner's successful completion of the Project. Consultant further acknowledges that it has reviewed the Schedule and that it is a reasonable schedule for the performance of Consultant's services. Consultant shall be responsible for utilizing whatever resources are necessary to complete its services within the duration(s) specified by the Schedule for the performance of Consultant's services, provided, however, that changes made by Owner to the dates such services are to start or to Consultant's scope of services shall lead to adjustments in the Schedule, if properly documented by Consultant. Consultant shall not be entitled to additional compensation for adjustments in the Schedule unless such adjustments extend the final date for submitting its last required deliverable at least ninety (90) days beyond the date shown on the Schedule at the time of the execution of the Agreement.

3.5   *Standard of Care* Consultant shall perform its services with the same care as would be used by a reasonable consultant performing similar services in the location of the project.

3.6   *Consultant's Project Manager* Consultant shall name a Project Manager who shall have full authority to carry out Consultant's responsibilities as specified in the Agreement.

## Article 4 Ownership of Work Product

4.1   *Consultant's Deliverables* Owner shall own Consultant's Deliverables produced by Consultant in the course of performing its services pursuant to the Agreement. Consultant shall deliver to Owner such Consultant's Deliverables and, if requested by Owner, copies of any studies, calculations, or other types of analyses (collectively, "Consultant's Analyses") on which such Consultant's Deliverables are based upon the termination or expiration of the Agreement. The delivery of all Consultant's Deliverables and, if requested by Owner, all Consultant's Analyses, shall be a precondition for final payment by Owner to Consultant under the Agreement.

4.2   *Limitation on Use* Consultant shall not use the Consultant's Deliverables for any purpose other than to provide its services pursuant to the Agreement without the advance written approval of Owner. If the Agreement is terminated prior to the completion of all of the Consultant's Deliverables, Owner shall be authorized to transmit Consultant's Deliverables and Consultant's Analyses such as they may be to a replacement Consultant and such replacement Consultant shall be authorized to use such Consultant's Deliverables and Consultant's Analyses to the full extent necessary to complete all Consultant's Deliverables. Owner shall not use Consultant's Deliverables for any purpose not associated with the Project without Consultant's written approval.

4.3   *Use by Owner Upon Termination* Termination by Owner pursuant to Paragraph 9.1, regardless of whether for cause or for convenience, shall not extinguish or limit in any way Owner's ownership of and right to use Consultant's Deliverables and Consultant's Analyses.

## Article 5 Payment

5.1   *Agreement Sum* As total compensation for all services to be provided by Consultant on the Project, Owner shall pay to Consultant the sum of _____ Dollars

($_____; "Agreement Sum"). The Agreement Sum is a lump sum amount which shall be modified only pursuant to the provisions of Article 6.

5.2    *Invoices* Consultant shall submit invoices on a monthly basis. The value of Consultant's invoices shall be calculated by multiplying the number of hours each person who performed services during the invoice period times each person's hourly rate, as set forth in *Attachment No. __*. Owner may request any documentation reasonably necessary to determine if Consultant is entitled to the amount set forth in Consultant's invoice. Owner shall make payment to Consultant within thirty (30) days of receipt of an acceptable invoice. Such payment shall be for the full amount of the invoice, unless the amount is disputed by Owner, in which case, Owner shall pay the undisputed amount within such thirty (30) days.

5.3    *Reimbursable Expenses* Owner shall reimburse Consultant for certain expenses to the extent such expenses are incurred by Consultant solely because of the provisions of its services pursuant to the terms of the Agreement ("Reimbursable Expenses"). Reimbursable Expenses shall include:

a.    travel related expenses including airfare, lodging, meals, car rentals, taxis;
b.    long distance telephone charges;
c.    messenger service and express mail;
d.    printing, reproduction, and plotting costs associated with reproducing the Consultant's Deliverables;
e.    fees and Reimbursable Expenses to be paid to subconsultants retained at the instruction of or with approval by Owner;
f.    services of any testing agencies retained at the instruction of or with the approval of Owner; and
g.    any other expense related to providing the services required by the Agreement and approved in advance by Owner.

## Article 6 Changes to the Agreement

6.1    *Amendments* Subject to Paragraph 6.3, the Agreement Sum, and any other terms of the Agreement, shall be modified only by written amendment ("Amendment") signed by both Owner and Consultant. No modification to any term of the Agreement shall take effect until the Amendment containing such modification has been fully executed.

6.2    *Additional Services* Increases, decreases, or other modifications to Consultant's Scope of Services shall be made only by Amendment to the Agreement. Services requested of Consultant by Owner which are not set forth in Article 3 and *Attachment No. __* shall be considered additions to the Scope of Services ("Additional Services") for which Consultant shall be entitled to an increase in the Agreement Sum. No increase in the Agreement Sum shall occur as a result of the performance of Additional Services unless Owner has given advance written approval for such Additional Services and such increase in the Agreement Sum.

6.3    *Unilateral Amendments* Notwithstanding any other provision of the Agreement, Owner may issue a unilateral (i.e., signed only by Owner) Amendment to Consultant modifying the Scope of Services and/or any other provision of the Agreement, and Consultant shall forthwith perform such Amendment. Such unilateral Amendment shall become binding as written on Consultant and Owner unless Consultant disputes such unilateral Amendment pursuant to the provisions of Article 10.

6.4    *Proposals for Additional Services* Any proposal for Additional Services submitted by Consultant, whether at Owner's request or at Consultant's instigation, shall set forth in appropriate detail the services to be provided, the value of the proposal, and the contractual basis on which Consultant believes it is entitled to an increase in the Agreement Sum. The value of proposals for Additional Services shall be established a) on the basis of a lump sum, or b) on the basis of time incurred, which method shall involve multiplying the number of hours to be worked by each person performing the services times the hourly rates of such persons, as set forth in *Attachment No. __.* Owner may request Consultant to submit any additional information or documentation Owner reasonably believes is necessary to determine the merit and/or appropriate value of the proposal.

6.5    *Not Eligible as Additional Service* Notwithstanding any other provision of this Agreement, no service performed by Consultant shall be an Additional Service if it is performed because of Consultant's negligence, error, or failure to perform services in accordance with the requirements of the Agreement.

## Article 7 Insurance

7.1    *Required Policies* Consultant shall at all times while performing services on the Project carry insurance policies at least sufficient to meet the following requirements with insurance firms authorized to provide such policies in the State of _____ and satisfactory to Owner:

   a.    Commercial General Liability Insurance, written on an "occurrence" basis, with limits of not less that $1,000,000 per occurrence and $2,000,000 in the aggregate.

   b.    In the event that motor vehicles will be used by the Consultant in the performance of its services, Automobile Liability Insurance at a limit of not less than $1,000,000 per accident.

   c.    Umbrella Liability Insurance at a limit of not less than $1,000,000 per occurrence and in the aggregate.

   d.    Professional/Errors and Omissions Liability Insurance at a limit of not less than $1,000,000 for each claim and in the aggregate.

7.2    *Named Insured* Each policy required by Paragraph 7.1 a–c shall name Owner as an additional insured.

7.3    *Certificates of Insurance* Consultant shall, prior to commencing the performance of services on this Project and thereafter as required by Owner, submit to Owner certificates of insurance which substantiate that Consultant has insurance policies in force which meet the requirements of Paragraph 7.1. Owner may require that any policy required by Paragraph 7.1 be submitted for its review and approval.

7.4    *Cancellation* Consultant shall notify Owner immediately if any policy required by the Agreement is cancelled for any reason, or if such policy is modified in any way the effect of which is that Consultant does not have insurance in force as required by Paragraph 7.1. Following the completion of its services on the Project,

Consultant shall not cancel any policy required by Paragraph 7.1 unless it has given Owner thirty (30) days advance notice in writing and such cancellation is permitted by Paragraph 7.1.

## Article 8 Indemnification

8.1    *Indemnification* To the fullest extent permitted by law, Consultant shall indemnify and hold harmless Owner and its officers, directors, and agents from and against claims, damages, losses and expenses, including but not limited to attorneys' fees, arising out of or resulting from performance of its Scope of Services, but only to the extent caused by the negligent acts or omissions of Consultant, a subconsultant, anyone directly or indirectly employed by them or anyone for whose acts they may be liable, regardless of whether or not such claim, damage, loss or expense is caused in part by a party indemnified hereunder.

## Article 9 Termination

9.1    *Termination* If Consultant fails to provide its services as required by the Agreement, or otherwise fails to comply with the requirements of the Agreement, Owner may terminate the Agreement by giving Consultant seven (7) days advance written notice. If Consultant does not correct such failure, or make significant and continuing progress in correcting failures which cannot be corrected within seven (7) days, Owner may terminate the Agreement at the end of such seven (7) day period. Owner may terminate the Agreement at any time for its convenience by giving Consultant seven (7) days advance written notice.

9.2    *Payment to Consultant* Upon termination, regardless of cause, Consultant shall submit within ten (10) days of the effective date of the termination an invoice for the value of the work performed up to the effective date of the termination. If Consultant was not terminated for any reason for which it bears responsibility, Consultant shall be permitted to seek reimbursement for reasonable direct costs solely attributable to such termination. Such reasonable direct costs shall not include anticipated revenues, lost profits, or any other anticipatory or consequential damages of any kind.

9.3    *Indemnification* If Consultant is terminated by Owner other than for Owner's convenience, Consultant shall hold Owner harmless for all damages and costs arising from Consultant's termination, including but not limited to reprocurement costs, loss of use and/or revenue because of delayed completion of the project, attorneys' and other professional services fees, and other costs.

## Article 10 Disputes

10.1    *Resolution of Disputes* All claims, disputes, and other matters in question between Owner and Consultant arising out of or relating to this Agreement or the breach thereof that cannot be resolved by the parties shall be submitted for resolution to a court of competent jurisdiction in the State of _____ unless otherwise agreed to by the parties. The parties hereby agree to negotiate in good faith any claims, disputes or other matters in question during the term of this Agreement before resorting to litigation.

10.2 *Claims* Consultant shall have the burden of documenting the merit and value of any claim it submits to Owner. Consultant shall continue to perform its Scope of Services notwithstanding the existence of any unresolved claim or dispute.

## Article 11 Records

11.1 *Consultant Shall Retain Records* Consultant shall retain copies of Consultant's Deliverables, Consultant's Analyses, memoranda, correspondence, submittals, and all other documents generated or received by Consultant in connection with its services under the Agreement (collectively, "Consultant's Project Records") for the term of the Agreement and for a period of six (6) years following the termination or expiration of the Agreement.

11.2 *Access to Consultant's Project Records* Owner may at any time during the term of the Agreement and for a period of six (6) years following the termination or expiration of the Agreement request access to and/or copies of any or all of Consultant's Project Records.

## Article 12 Miscellaneous Provisions

12.1 *Calendar Days* References to days in the Agreement shall mean calendar days unless otherwise stated. When a specified period would end on a Saturday, Sunday, or holiday, it will be deemed to end on the next business day.

12.2 *Laws of* _____ The provisions of the Agreement shall be construed according to the laws of the State of _____.

12.3 *Headings* The headings contained in the Agreement are for convenience only; they are not part of the Agreement and shall not define or limit the scope of any provision of the Agreement.

12.4 *Provision Unenforceable* If any provision of the Agreement is held to be unenforceable by a court of competent jurisdiction, the remainder of the Agreement shall remain in effect as if the invalid term had not been initially included.

12.5 *Conflict of Interest* Consultant, by executing the Agreement, certifies that it has no current or known future obligation to perform services that would constitute a conflict of interest with its obligations to Owner under the Agreement.

12.6 *No Third Party Rights* The Agreement confers no rights or benefits on persons or entities that are not signatory to the Agreement, it being agreed that there are no intended third party beneficiaries to the Agreement.

12.7 *Limitation of Liability* No officer, director, shareholder, partner or agent of Owner, nor any person or entity holding any interest in Owner shall be personally liable, whether directly or indirectly, by reason of any default by Owner in the performance of any of the obligations under the Agreement, including without limitation, Owner's failure to pay Consultant as required hereunder.

12.8 *Notices* Written notices required by the Agreement to be sent by one party to the other shall be sent to the addresses first appearing in the Agreement to the attention of the receiving party's Project Manager. Either party may change the contractual notice address by providing written notice to the other party stating the new notice address.

12.9 *Assignment* Consultant agrees that Owner may assign the Agreement to any affiliate, equity investor, lender, or any other entity to which Owner may in its reasonable discretion elect to assign the Agreement. Such assignment shall not be the basis for any request for additional compensation or any adjustment to the Schedule.

Owner and Consultant each execute the Agreement under Seal as of the date first mentioned in the Agreement to fully memorialize the agreements and mutual consideration contained in the Agreement. The Agreement shall be executed in multiple copies, each of which shall be considered an original.

| Owner: | | Consultant: | |
|---|---|---|---|
| By: | _____ | By: | _____ |
| Title: | _____ | Title: | _____ |

# Appendix C
## Construction contract

This contract ("Contract"), made this _____ day of _____, 20__, for valuable consideration by and between _____ with a principal place of business at _____ ("Owner") and _____ with a principal place of business at _____ ("Contractor") for the construction of _____ at _____ ("Project").

## Article 1 The Contract Documents

1.1 *Contract Documents* The Contract Documents shall consist of this Contract, the plans and specifications listed in Exhibit A, and any other Exhibits enumerated in Paragraph 1.3, any written Change Orders executed pursuant to Article 9 after the effective date of this Contract, and _____ [any other documents the owner believes should be Contract Documents].

1.2 *[Applicable State] Law Incorporated* All provisions of [applicable state] law required to be included in the Contract Documents including statutes, rules, regulations, ordinances, orders, or any other provision of law are hereby incorporated by reference in the Contract Documents and shall be considered a part of the Contract Documents as if they were written out completely whether or not they are in fact so written.

1.3 *Exhibits* The following Exhibits are hereby incorporated in and made a part of the Contract Documents:

- Exhibit A: Plans and Specifications ("Design Documents");
- Exhibit B;
- Exhibit C;
- [Additional exhibits if necessary].

1.4 *Ambiguities and Contradictions* If there is any ambiguity, inconsistency, or error in or among the Contract Documents, such ambiguity, inconsistency, or error shall be resolved on the basis of the priority reasonably determined by Owner as being consistent with the overall intent of the Contract Documents and required to produce the intended result. Subject to such determination by Owner, the Contract Documents shall be interpreted on the basis of the following priorities:

a. Change Orders,
b. The Contract,
c. Specifications,

   d.   Plans, and
   e.   Other exhibits other than Exhibit A, if any.

   If such ambiguity, inconsistency, or error occurs within the same or similar documents, the issue shall be resolved so as to provide Owner with the higher quality and/or greater quantity of materials, equipment, and/or labor.

## Article 2 The Scope of the Work

2.1   *The Work* Contractor shall perform all of the work required by the Contract Documents ("the Work"). The Work shall include all the construction and services required by the Contract Documents and all other labor, materials, equipment, management, coordination, and other services to be provided by Contractor to fulfill Contractor's obligations. It is understood that the Contract Documents provide for a complete Project, and Contractor shall perform all work necessary and reasonably inferable from the Contract Documents to provide such complete Project for the Contract Price.

## Article 3 Contractor's Responsibilities

3.1   *Review Contract Documents* Contractor shall carefully review the Contract Documents prior to submitting its bid. By submitting a bid, Contractor represents that it has conducted such a review, and that the Contract Documents contain no ambiguities or other problems which would preclude Contractor from performing the Work as required by the Contract Documents.

3.2   *Site Visit* Contractor shall visit the location described in the Contract Documents where the Project shall be built ("Site") to determine the conditions of the Site. To the extent that any condition could reasonably have been observed by such a visit, such condition shall not serve as a basis for a request by Contractor for an increase in the Contract Price or an extension of the Contract Time.

3.3   *Subcontracting* To the maximum extent permitted by law, Contractor may subcontract the Work to the extent and in whatever manner it deems most appropriate to complete the Work as required by the Contract Documents. Contractor shall not employ any Subcontractor to whom Owner has reasonable objection, either initially or as a substitute. Contractor shall not be required to employ any Subcontractor against whom it has a reasonable objection.

3.4   *Subcontractors and Subcontracts* Any organization, person, or entity furnishing labor, services, materials, and/or equipment to Contractor shall be a Subcontractor for the purposes of this Contract, regardless of whether it has a Subcontract with Contractor or another Subcontractor, and regardless of whether it provides labor or services or only provides materials or equipment. Subcontractors shall only furnish labor, services, materials, and/or equipment pursuant to a written agreement approved by Owner. All Subcontracts shall specifically make all applicable provisions of the Contract Documents binding requirements on such Subcontractor for the benefit of Owner. Contractor shall ensure that no Subcontractor commences work at the site prior to the full execution of a Subcontract, unless the commencement of such work is approved in writing by Owner.

3.5   *All Necessary Resources* Contractor shall provide all labor, services, materials, equipment, management, and other resources necessary to complete the Work as required by the Contract Documents.

3.6   *Work Must Comply with Contract Documents* The Work performed by Contractor shall comply with all requirements of the Contract Documents. No provision of the Contract Documents shall relieve Contractor from such obligation. Neither inspection by Owner, Designer, or any party authorized pursuant to Paragraph 7.3 to act on Owner's behalf, nor payment by Owner, shall relieve Contractor of the obligation to perform the Work as required by the Contract Documents.

3.7   *Contractor's Project Manager* Contractor shall appoint a Project Manager with full authority to carry out all of Contractor's responsibilities under this Contract and a Superintendent who shall manage Contractor's operations at the Site. The Project Manager and the Superintendent shall have the experience and skills necessary to manage Contractor's performance of the Work. Owner may, in its sole discretion, approve Contractor's Project Manager and Superintendent, and require Contractor to replace the Project Manager and/or the Superintendent.

3.8   *Contractor's Personnel* Contractor and all its Subcontractors shall provide competent, qualified, and reliable personnel to perform the Work. Contractor shall at all times maintain good working order at the Site. Owner reserves the right to preclude any person from working on the Project.

3.9   *Comply with Applicable Laws* Contractor shall become familiar with and shall comply with all applicable provisions of federal, state, and local laws, including rules and regulations.

3.10  *Coordination of the Work* Contractor shall be fully responsible for the coordination and management of all of the Work, including subcontracted work. Disputes between Subcontractors and/or between Contractor and any Subcontractor relating to responsibility for performing portions of the Work shall be resolved by Contractor at no additional cost to Owner, except to the extent that such disputes result directly from deficiencies in the plans and/or specifications which could not have been identified prior to commencement of the Work.

3.11  *Manufactured Materials* All manufactured materials shall be handled, stored, installed, cleaned, and protected in accordance with the manufacturer's directions unless otherwise indicated in the Contract Documents. Contractor at time of Substantial Completion shall assign to Owner all manufacturer's warranties required by the Contract Documents and shall perform the Work in such manner as to preserve all such manufacturer's warranties.

3.12  *Responsible for Subcontractors' Acts and Omissions* Contractor shall be fully responsible for all the acts and omissions of all of its Subcontractors at whatever tier.

3.13  *Contractor's Schedule* Contractor shall prepare, submit, and update a project schedule in accordance with the requirements of this Paragraph 3.13. Within fifteen (15) days following execution of this Contract, or by a later date if approved in writing by Owner, Contractor shall submit an initial schedule for approval by Owner. The submission and approval of such initial schedule shall be a precondition to Owner making any payment pursuant to this Contract. Thereafter, Contractor shall submit an update of such schedule monthly for approval as part of its requisition for payment until final completion, or until Owner relieves Contractor of this responsibility in writing. Failure to submit a schedule update by Contractor, or to obtain Owner's approval for such update, may serve as a basis for Owner to deem the associated requisition for payment incomplete and not ready for processing until accompanied by such approved schedule update. Unless another format is approved in writing by Owner, all schedules required by this Paragraph shall be critical path method ("CPM") schedules. If Contractor falls behind the most recently approved schedule for reasons that are its responsibility, Contractor shall submit a recovery schedule for approval by Owner and shall implement such recovery schedule at no additional cost to Owner. A written instruction by Owner to Contractor directing it to submit and implement such recovery schedule shall not constitute an instruction to accelerate. Approvals required by this Paragraph 3.13 shall be limited to determining that any schedule submission provides a realistic approach to completing the Project within the Contract Time and that any update also accurately reflects the progress of the Work.

3.14 *Contractor to Keep Working* Contractor shall at all times prosecute the Work so as to complete the work by the contractually specified completion date. Contractor shall continue to perform the Work notwithstanding the existence of any outstanding change order proposals or claims, and Contractor shall not suspend or stop work for any other reason than a written instruction by Owner issued as provided by Paragraphs 7.6 or 7.7 or Article 12.

3.15 *Coordination with Other Contracts* If Owner elects to award portions of the Work, or other work, to one or more other contractors, Contractor shall coordinate its Work with such other contractors. If Contractor's Work is adversely impacted by the activities of such other contractor, Contractor shall seek appropriate reimbursement from such other contractor.

3.16 *Indemnification* To the greatest extent permitted by law, Contractor shall hold Owner harmless and shall pay all damages and all reasonable costs, including but not limited to attorney, consultant, and other professional service fees incurred by Owner to the extent such damages and costs are paid or incurred by Owner as a result of Contractor's failure to perform the Work as required by the Contract Documents or as a result of bodily injury and/or property damage caused by Contractor's negligence. Contractor's responsibilities under this Paragraph shall include all acts and omissions of its Subcontractors.

3.17 *As-Built Design Documents* Contractor shall maintain a fully conformed set of the Design Documents for the Project which shall be available for Owner's and Designer's inspection at any time. Upon completion of the Work, Contractor shall submit to Designer for Designer's approval a complete set of the as-built Design Documents.

3.18 *Safety* Contractor shall be responsible for all aspects of safety on the project. It shall be Contractor's responsibility to avoid, to the maximum extent feasible, bodily injuries and property damage resulting from performance of the Work. At least thirty (30) days prior to commencing work at the Site, Contractor shall submit to Owner for Owner's approval, a safety plan indicating how Contractor shall maximize safety on the Project, and Contractor shall not commence work at the Site until such safety plan is approved in writing by Owner. Owner's approval for the purposes of this Paragraph 3.18 shall be limited to determining that Contractor's safety plan represents a reasonable approach to ensuring safety on the project. Contractor shall designate a key member of its project team as Safety Manager. The Safety Manager shall be responsible for implementing Contractor's safety plan.

3.19 *Quality Control* Prior to commencement of any work, Contractor shall submit to Owner for Owner's approval a quality control plan indicating how Contractor intends to ensure that the Work complies with the requirements of the Contract Documents. Contractor shall not commence work at the Site until such quality control plan has been approved in writing by Owner. Contractor's quality control plan shall include the identification of a member of Contractor's staff who shall be responsible for the implementation of the plan. Owner's approval for the purposes of this Paragraph 3.19 shall be limited to a determination that the plan represents a reasonable approach to ensuring the Work complies with the Contract Documents.

## Article 4 The Term of the Contract

4.1 *Contract Time* Contractor shall substantially complete the Work within _____ days following the date specified in a written Notice to Proceed ("NTP") to be issued by Owner to Contractor ("Contract Time"). The Contract Time shall be adjusted only pursuant to Article 9. Contractor shall not commence work until Owner issues an NTP. Owner's NTP may apply to all or a portion of the Work.

4.2 *Substantial Completion* The Work shall be deemed to be substantially complete when Contractor has substantially completed the Work and Owner determines that it is able to occupy and use the Project for its intended purpose(s) notwithstanding that minor amounts of the Work remain to be completed or corrected ("Substantial Completion"). Owner's ability to occupy and use the facility shall mean that all building systems are fully operational, that the facility is readily useable by Owner for its intended purpose(s), that Contractor has obtained a certificate of occupancy or, with Owner's written approval, a temporary certificate of occupancy; and that any other requirements imposed by any authorities having jurisdiction have been satisfied. Notwithstanding any other provision of this Paragraph or of the Contract, Substantial Completion shall not occur until granted in writing by Owner.

4.3 *Substantial Completion Certificate* Upon Substantial Completion, Owner shall issue a certificate establishing that the Project has achieved Substantial Completion ("Substantial Completion Certificate"). The Substantial Completion Certificate shall document agreement between Owner and Contractor concerning a) the portion of the Work remaining to be completed or corrected ("Substantial Completion Work"); b) the value of the Substantial Completion Work; c) the schedule for completion of the Substantial Completion Work; d) the protection of the Substantial Completion Work, including applicable insurance coverage; and e) any other relevant conditions related to the Substantial Completion Work.

4.4 *Time Is of the Essence* Contractor hereby acknowledges its understanding that time is of the essence of this Contract, including periods of time for notices, submissions, and all other contractually required actions. Contractor further acknowledges its understanding that Contractor's failure to complete the Project within the Contract Time may well cause Owner to incur significant additional costs for which Contractor, to the extent it is responsible for such costs, shall be required to reimburse Owner.

## Article 5 The Contract Price

5.1 *Contract Price* Contractor shall receive _____ Dollars ($_____) for the full and complete performance of the Work ("Contract Price"). The Contract Price shall only be increased or decreased pursuant to Article 9.

## Article 6 Payment

6.1 *Schedule of Values* Contractor shall submit, for approval by Owner, a schedule of values which shall break the Scope of Work into appropriate specific activities, provide a value for each activity such that the total of the activity values shall at all times equal the Contract Price; and provide columns that show for each activity, respectively, the original value of the activity, the value of the work performed to date, the value of the work performed during the current period, the percentage of the Work completed to date, and the value of the balance of the work to be performed ("Schedule of Values") The submission and approval of the initial Schedule of Values shall be a precondition to Owner making any payment to Contractor pursuant to this Contract. Contractor shall submit a copy of the Schedule of Values with each requisition for payment which shall reflect the most current values at the time the Schedule of Values is submitted.

6.2 *Requisition for Payment* Contractor shall submit to Owner its requisition for payment ("Requisition") on or before the fifth (5th) day of each month, or such other day as Owner and Contractor shall agree upon, for work performed the previous month. Such Requisition shall include a calculation showing the Contract Price as stated in Paragraph 5.1, the value of all Change Orders, and the resulting current Contract Price; and a calculation showing the value of all work performed through the last day of the previous month, the total of all previous payments to Contractor, and the resulting total due for that Requisition. Contractor shall submit with each Requisition evidence satisfactory to Owner that all amounts owed to Subcontractors prior to the submission of such Requisition have been paid to such Subcontractors. Contractor shall submit with each Requisition all documentation reasonably required by Owner to substantiate the amount requested by such Requisition.

6.3 *Progress Payments* Owner shall approve each Requisition. Within thirty (30) days after Owner approves a Requisition requesting payment of the amount due for the preceding month, Owner shall make a progress payment to Contractor in an amount equal to the amount requested by the Requisition less (1) the value of any disputed amounts and any other claims by Owner against Contractor and (2) a retention of five percent (5%) of the approved amount of the Requisition. If Owner disputes any portion of a Requisition and withholds any portion of the amount claimed in such Requisition, Owner shall pay such withheld amount to Contractor within three (3) days of the Owner and Contractor resolving such disputed portion of a Requisition.

6.4 *Payment for Materials* Contractor may include in any Requisition the value of materials not incorporated in the Work but delivered and suitably stored at the Site (or at some other location approved by Owner in writing) to which Contractor has title or to which a Subcontractor has title and has authorized Contractor to transfer title to Owner.

6.5 *Costs Not Allowed* The Contractor shall not in any event be entitled to reimbursement of the following costs:

a. the salaries and associated payroll costs of any personnel not assigned to the Project unless approved in writing by Owner;

b. home office overhead or any costs associated with any offices other than the office located at the Site;

c. costs of tools or equipment with a value of less than five hundred dollars ($500) when new;

d. any costs which arise from the performance of defective work or from any other failure to perform the Work according to the requirements of the Contract Documents;

e.  interest or any costs associated with the use of capital for the performance of the Work;

f.  any differences between anticipated and actual profits or revenue, whether associated with the performance of the Work or not;

g.  any costs of consultants or attorneys used to prepare change order proposals or claims; and

h.  any other costs not specifically attributable to the performance of the Work.

6.6  *Withholding Payment* Owner may withhold all or a portion of a progress payment to Contractor if Owner reasonably believes that Contractor has failed in one or more significant ways to perform the Work as required by the Contract Documents; has failed to make timely payments to Subcontractors as required by Paragraph 6.7 below; or has breached this Contract in any other material manner.

6.7  *Progress Payments to Subcontractors* Within five (5) business days after Contractor receives a progress payment, Contractor shall pay to each Subcontractor the amount owed to such Subcontractor for work performed and not previously paid less any amount claimed due from the Subcontractor by Contractor. Contractor shall make a reasonable effort to resolve any disputed amounts promptly and shall pay any agreed upon amounts within three (3) days of reaching such agreement. Owner may, but shall not have any obligation to, withhold any amounts disputed between Contractor and any of its Subcontractors from its payment to Contractor as provided in Paragraph 6.3.

6.8  *Payment Following Substantial Completion* Upon issuance of the Substantial Completion Certificate by Owner, Contractor may submit a Requisition for the entire balance of the Contract Price not previously paid to Contractor less a) an amount equal to two hundred percent (200%) of the value of the Substantial Completion Work and b) the value of any other claims by Owner against Contractor. Such Requisition shall be submitted and paid in accordance with the provisions of this Article 6.

6.9  *Conditions for Final Payment* Owner shall not make the final payment to Contractor until Contractor has:

a.  finally completed all of the Work;

b.  submitted to Designer, and Designer has approved, a set of as-built Design Documents;

c.  submitted to Owner, and Owner has approved, all warranties required by the Contract Documents;

d.  submitted evidence satisfactory to Owner that Contractor has paid all amounts owed to Subcontractors, or to the extent such payments have not been made, Contractor has provided Owner with a justification therefore satisfactory to Owner; and

e.  satisfied all other obligations imposed by the Contract.

6.10 *Final Payment* Final payment shall be requested by Contractor and made by Owner in accordance with the provisions of this Article 6. Acceptance of such final payment by Contractor shall result in the waiver of all claims by Contractor against Owner except those identified by Contractor in writing no more than twenty-one (21) days prior to accepting final payment.

## Article 7 Owner's Responsibilities

7.1    *Owner's Project Manager* Owner shall appoint a Project Manager with full authority to carry out all of Owner's responsibilities under this Contract. Owner may at any time change the Project Manager by sending Contractor written notice providing the name of the new Project Manager and the effective date of the change.

7.2    *Designer* Owner has retained the services of _____ ("Designer") to create the plans and specifications referenced in Paragraphs 1.1 and 1.3. Contractor shall cooperate with Designer as necessary to fulfill its responsibilities under this Contract, provided that Contractor shall not perform any design services except as required by the Design Documents and shall not be responsible for any aspect of Designer's work.

7.3    *Additional Consultants* Owner may elect to retain a program or construction manager, an owner's representative, a clerk of the works, and/or any other consultants to assist in the management of this Project. The utilization of any such person or organization shall not change the responsibilities of Contractor under this Contract.

7.4    *Access to Site* Owner shall provide Contractor access to the Site on the date stated in the NTP described in Paragraph 4.1. Such access shall be complete and unimpeded in any fashion, except as provided in the Contract Documents.

7.5    *Owner May Award Work to Separate Contractors* Pursuant to the provisions of Article 9, Owner may remove portions of the Work from this Contract and may award such portions to one or more other contractors. Owner may award other work associated with the Project to contractors other than Contractor.

7.6    *Owner May Stop Work* Owner may order Contractor to stop the Work, or any portion of the Work, if Contractor materially or persistently fails to perform the Work as required by the Contract Documents. Such order shall be in writing and shall remain in effect until Owner provides written authorization for the stopped work to proceed. Such an order to stop all or a portion of the Work shall not result in an increase in the Contract Price or an extension of Contract Time unless Contractor demonstrates that the stopped work did comply with all applicable contractual requirements. The authority to issue an order to stop the Work shall not, under any circumstances, create a duty on the part of Owner to issue such an order.

7.7    *Owner May Suspend Work* Owner may order Contractor to suspend or delay all or any part of the Work for such period of time as it may determine to be appropriate for the convenience of Owner. Such order shall be in writing and shall remain in effect until rescinded in writing by Owner. If such suspension or delay exceeds sixty (60) days, Contractor shall be entitled to seek an increase in the Contract Price and/or an extension of the Contract Time to the extent it can demonstrate an increase in costs and/or a need for additional time solely attributable to such suspension or delay.

7.8    *Correction of Defective Work* Owner may correct any work which does not comply with the requirements of the Contract Documents ("Defective Work"), using its own forces or another contractor, provided it has notified Contractor in writing of the Defective Work, and Contractor has not, within seven (7) days, corrected such defective work, or in the case of work requiring more than seven (7) days to correct, taken all possible steps within seven (7) days to begin the correction and to ensure its completion at the earliest possible moment. The cost of such corrective action, including the cost of any necessary design, project management, or other consulting services, shall be the responsibility of Contractor. Owner may deduct such cost from any money owed to Contractor, or, if insufficient amounts are owed to Contractor, Contractor shall pay any balance to Owner.

## Article 8 Administration of the Contract

8.1 *Inspection of the Work* Designer and Owner shall inspect the progress of the Work. Any such inspection activity by Designer and/or Owner shall not in any way reduce Contractor's obligation to perform the Work as required by the Contract Documents.

8.2 *Identification of Defective Work* Designer and Owner shall, as part of any inspections of the Work, identify Defective Work and shall notify Contractor in writing of such Defective Work. Such notices shall be on a defective work report form issued or approved by Owner and shall be the subject of a defective work log administered by Owner.

8.3 *Review of Requisitions* Owner and, if requested by Owner, Designer shall review each Requisition submitted by Contractor. Owner shall determine what action to take with respect to such Requisition. Such review shall consider the accuracy of the supporting documentation, including but not limited to, the amounts shown in the Schedule of Values required to be submitted with the Requisition by the Contract; the amount of the Work performed and the extent to which it complies with the requirements of the Contract; and the mathematical accuracy of the Requisition. The monthly update schedule required by Paragraph 3.13 and the Schedule of Values required by Paragraph 6.1 shall be submitted with the Requisition, and the absence of such update schedule or Schedule of Values shall make the Requisition incomplete and may, in Owner's discretion, serve as a basis for not approving the Requisition as required by Paragraph 6.3.

8.4 *Submittals* Contractor shall submit all shop drawings, product samples, and other submittals required by the Contract Documents to Designer. Designer shall review all such shop drawings, product samples, and other submittals by Contractor required by the Contract Documents in sufficient detail to confirm that such shop drawings, product samples, and other submittals comply with all applicable provisions of the Contract Documents. Designer shall mark each such shop drawing, product sample, or other submittal as either "Approved," "Approved as noted," "Resubmit with changes as indicated," or "Rejected." Designer may substitute its standard wording for those in the previous sentence, if approved in writing by Owner, provided that Designer shall agree that whatever words are substituted for "Approved" shall have the legal effect of approval. If Contractor performs work depicted by shop drawings, involving the installation of products, or which depends on other submittals required by the Contract Documents prior to the approval by Designer of such shop drawings, product samples, or other submittals required by the Contract Documents, Contractor performs such work at its own risk. Approval by Designer of any shop drawing, product sample, or other submittal required by the Contract Documents shall not extend to any deviation from the requirements of the Contract Documents unless Contractor provides advance written notice of such deviation.

8.5 *Requests for Substitution* Contractor may submit a request for substitution ("Request") in connection with any specified material or equipment. The Request shall be submitted to Designer in writing at least thirty (30) days prior to the commencement of any construction work on the Project. The Request shall only be approved if the substitution would not directly or indirectly cause an increase in the Contract Price or an extension in the Contract Time and if Designer determines that the proposed material or equipment would perform in substantially the same manner as the originally specified material or equipment and that the substitution would benefit the Project. Owner, in its sole discretion, may instruct Designer to approve or reject the Request regardless of whether it meets the requirements of this Paragraph. Designer shall issue a written decision within twenty-one (21) days of receipt of the Request.

## Article 9 Changes in the Work

9.1   *Owner May Order Changes* Owner shall be authorized, without giving notice to Contractor's surety, to make changes in the Work consistent with the intent of the Project, including additions, deletions, and/or modifications to the Scope of the Work. Such changes may involve changes in the Contract Price and/or the Contract Time.

9.2   *Written Change Orders Required* Changes in the Contract Price and/or the Contract Time shall only be made by written change order ("Change Order") pursuant to the terms of this Article 9. A Change Order shall become fully binding on Contractor when it is executed by both Owner and Contractor; provided that Owner may issue a Change Order executed only by Owner ("Unilateral Change Order") and such Unilateral Change Order shall become fully binding as a Change Order upon Contractor and Owner unless it files a notice of claim within five (5) days of receipt of such Unilateral Change Order and otherwise complies with Article 11.

9.3   *Owner Requested Proposals* Owner may, in connection with any change in the Work made at its instruction, request a proposal from Contractor for the extra costs and/or time associated with the performance of such change, if any. Following the receipt of such a request, Contractor shall have twenty-one (21) days in which to submit its proposal, except that Owner may extend such period by written notice to Contractor. Failure by Contractor to submit its proposal within the twenty-one (21) days, or when applicable, the extended period of time, shall constitute a complete waiver by Contractor of any right to seek an increase in the Contract Price and/or an extension in the Contract Time associated with the subject of the invitation for proposal.

9.4   *Contractor's Notice and Proposal* Contractor shall, within five (5) days of the happening of an act, omission, or circumstance which it believes entitles it to an increase in the Contract Price and/or an increase in the Contract Time, submit a written notice to Owner of its intent to submit a change order proposal. Contractor shall have twenty-one (21) days from the date of its written notice to submit its change order proposal to Owner, except that Owner may extend such period by written notice to Contractor. Failure by Contractor to submit such notice within five (5) days following the happening of such act, omission, or circumstance, or failure to submit such proposal within twenty-one (21) days, or when applicable, the extended period of time, following submission of its notice shall constitute a complete waiver by Contractor of its right to seek any increase in the Contract Price or any extension of the Contract Time in connection with such act, omission, or circumstance.

9.5   *Differing Site Conditions* If, during the progress of the Work, Contractor discovers that physical conditions encountered at the Site differ substantially from those shown on the plans or indicated in the Contract Documents, Contractor may request an adjustment in the Contract Price and/or the Contract Time as provided in this Article 9.

9.6   *Sufficient Documentation* Each change order proposal submitted pursuant to Article 9 shall be submitted by Contractor to Owner and, at Owner's request, with a copy to Designer, and shall be accompanied by documentation sufficient to substantiate Contractor's proposal. Proposals seeking an extension in the Contract Time shall include a schedule analysis substantiating the extension sought in the proposal. Owner may request additional documentation if in its judgment such additional

documentation is reasonably necessary to determine if Contractor's proposal is justified as to entitlement and/or value.

9.7 *Change Order Pricing Methods* Every Change Order shall be priced using one or more
of the methods set forth in this Paragraph. The price of each Change Order shall be
based on a lump sum amount agreed to by Owner and Contractor; unit prices included
in the Contract or agreed to by Owner and Contractor; or the actual cost of the work
involved in the Change Order as demonstrated by documentation submitted by
Contractor and approved by Owner.

9.8 *Cost of the Work* When determining the actual cost of the work involved in a
Change Order ("Change Order Work"), such cost shall be limited to the costs of per-
forming the Change Order Work, whether performed by Contractor or one or more
Subcontractors, including:

  a.  payroll costs for workers and first line supervisors, including wages and the ben-
      efits normally paid by Contractor or Subcontractor to its employees, plus the
      burdens required by law and/or by applicable collective bargaining agreements,
      but in the case of payments called for by collective bargaining agreements, only to
      the extent such burdens are for benefits payable or collectable by individuals;
  b.  payments to suppliers of materials and/or equipment to be included in the Work,
      including any transportation and storage costs;
  c.  payments, not to exceed rates set forth in the Blue Book published by Dataquest,
      for the lease or rental of equipment necessary to perform the Change Order Work,
      including standby time necessitated by reasons outside of Contractor's control, or
      equivalent costs for the use of Contractor owned equipment based on Contractor's
      standard accounting practices;
  d.  payments for additional premiums for bonds and/or insurance necessitated by the
      Change Order Work;
  e.  payments for any other costs solely attributable to the performance of the Change
      Order Work and not precluded by Paragraph 6.5; and
  f.  the mark-ups authorized by Paragraph 9.10.

9.9 *Work Documented by Daily Reports* When establishing the price of a Change Order
by determining the actual cost of the Change Order Work, Contractor shall submit
daily reports to Owner setting forth which workers and first line supervisors performed
the Change Order Work, including their names, trades, and how many hours each
worked; the materials and/or equipment installed that day; the equipment utilized to
perform the Change Order Work that day; and an itemization of any other costs incurred
that day which Contractor believes it is entitled to include in the value of the Change
Order pursuant to Paragraph 9.8. Such daily reports shall constitute substantiation of
the cost of the Work only if signed by Contractor and by Owner indicating Owner's
agreement with the information on the daily report.

9.10 *Overhead and Profit* When establishing the price of a Change Order, Contractor shall
be entitled to a fee covering overhead and profit not to exceed fifteen percent (15%) of
the cost of the work actually performed by Contractor. When the Change Order Work
is performed by a first tier Subcontractor, that Subcontractor shall be entitled to a fee
not to exceed fifteen percent (15%) of its total costs, and Contractor shall be entitled
to a fee not to a exceed five percent (5%) of the Subcontractor's costs excluding
the Subcontractor's fee. When the Change Order Work is performed by a lower tier

Subcontractor, that Subcontractor shall be entitled to a fee not to exceed fifteen percent (15%) of its total costs, and Contractor and the Subcontractor with whom the performing Subcontractor has contracted shall be entitled to a combined fee to be distributed in Contractor's discretion not to exceed five percent (5%) of the lower tier Subcontractor's total costs, excluding the lower tier Subcontractor's fee. In no event shall the amount to be paid for overhead and profit on any Change Order exceed twenty percent (20%) of the costs of performing the work.

9.11 *Time Extensions* Contractor may seek an extension in the Contract Time whenever the performance of the Work is delayed by any act or omission of Owner, by changes ordered in the Work, by the work of other contractors or other organizations employed by Owner, by differing site conditions, or by any other causes beyond the control of Contractor. When seeking a time extension under any of these circumstances, Contractor shall demonstrate that there is a need for a time extension notwithstanding Contractor's reasonable efforts to avoid or mitigate the need for such extension; and that the extension is necessary to perform those activities controlling the timing of the completion of the Work as demonstrated by Contractor's schedule required by Paragraph 3.13. Contractor shall not be entitled to an extension in Contract Time unless all of the Total Float (i.e., the total of the differences between early completion and late completion of the tasks on the critical path) has been consumed and the additional time to perform such activities necessarily extends the Contract Time. If Owner has approved a schedule format other than a CPM schedule pursuant to Paragraph 3.13, Contractor shall be entitled to a time extension only if it meets all the requirements of this Paragraph except that relating to Total Float.

9.12 *Remedies for Delay* A time extension shall be Contractor's sole and exclusive remedy for any delay of whatever kind and however caused except for those delays directly caused by differing site conditions, changes in the Work ordered by Owner, or acts or omissions of Owner. Contractor shall be entitled to an increase in the Contract Price under this Paragraph only to the extent of increased costs directly caused by such differing site conditions, changes in the Work, or acts or omissions of Owner.

9.13 *Change Order Full Compensation* The increase in Contract Price and/or extension in Contract Time included in any Change Order shall be Contractor's full compensation, including all direct, indirect, supplemental, and all other costs of any kind, for the acts, omissions, or circumstances giving rise to such Change Order, and Contractor shall not seek any additional compensation under Article 9 or Article 11 for such acts, omissions, or circumstances.

## Article 10 Insurance and Bonds

10.1 *Insurance Coverages* Contractor shall purchase and maintain until the Work is finally completed the following insurance coverages from companies authorized to issue such insurance policies in [state in which project is located] for not less than the following amounts:

    a.   Workers Compensation Insurance, as necessary to meet its obligations under the laws of [state in which project is located];

    b.   Employers Liability Insurance with limits of not less than $_____ for bodily injury by accident or disease;

    c.   Commercial General Liability Insurance with limits of not less than $_____ combined single limit for bodily injury and property damage liability;

d. Automobile Liability Insurance covering the ownership, maintenance, or use of all owned, non-owned, and hired vehicles used in the performance of the Work, including loading and unloading, with limits of not less than $_____ combined single limit for bodily injury and property damage liability; and

e. Umbrella Liability Insurance with limits of not less than $_____ which will provide bodily injury, personal injury, and property damage liability at least as broad as the primary coverages set forth above, including employer's liability and commercial general liability insurance.

10.2 *Commercial Liability Insurance* Contractor's Commercial General Liability Insurance shall include all products, premises-operations and completed operations liability (for at least five (5) years following Final Completion), independent contractors, and blanket contractual liability insurance covering all liabilities assumed under this Contract.

10.3 *Builder's Risk Insurance* Owner shall purchase and maintain Builder's Risk Insurance. This insurance shall cover the Work, including all materials and equipment to be incorporated in the Work, but not including tools, construction equipment, materials or supplies not included in the Work, and temporary structures or other property owned or rented by Contractor. Such insurance shall cover direct physical loss or damage caused by such events as fire, wind, explosion, vandalism, earthquakes, and flood.

10.4 *Certificates of Insurance* Contractor shall not commence or continue to perform the Work unless certificates have been submitted to Owner evidencing that all required insurance coverages are currently in full force and effect. Owner may withhold any progress payment otherwise due to Contractor if any such certificate has not been submitted when required.

10.5 *Cancellation* Contractor shall not permit any coverages required by Paragraph 10.1 to expire until the Work is complete, and Contractor has provided Owner written notice at least thirty (30) days in advance of the cancellation date.

10.6 *Bonds* Contractor, prior to commencing the Work, shall furnish performance and payment bonds, each in the amount of the Contract Price stated in Paragraph 5.1. Such bonds shall be furnished by a surety company licensed to issue such bonds by [state in which project is located] and shall be in form and content satisfactory to Owner.

## Article 11 Claims and Disputes

11.1 *Claims* A claim ("Claim") is a demand for an adjustment in the Contract Price and/or the Contract Time arising out of or related to the performance of this Contract. All Claims shall be submitted and decided solely pursuant to the terms of this Article 11. Contractor shall bear the burden of proof with respect to both entitlement and the value of any adjustment to the Contract Price and/or the Contract Time.

11.2 *Claims Must be Change Order Proposals* Unless approved by Owner in writing, no Claim shall be submitted unless the act, omission, or circumstance giving rise to the Claim shall have first been submitted as a change order proposal and acted upon pursuant to the provisions of Article 9.

11.3 *Submission of Notice and Claim* Contractor shall submit a notice of its intent to submit a Claim to Owner, and at Owner's request, with a copy to Designer, within five (5) days of the act, omission, or circumstance giving rise to the Claim. Contractor shall have twenty-one (21) days from the date of its notice to submit its Claim, except that Owner may extend such period by written notice to Contractor. Failure to submit such notice within five (5) days following the happening of such act, omission, or circumstance, or failure to submit such Claim within twenty-one (21) days, or, when applicable, the extended period of time, following submission of its notice shall constitute a complete waiver by Contractor of its right to seek any increase in the Contract Price or any extension of the Contract Time in connection with such act, omission, or circumstance.

11.4   *Sufficient Documentation* Each Claim submitted by Contractor shall be submitted to Owner, and at Owner's request, with a copy to Designer, and shall be accompanied by documentation sufficient to substantiate both Contractor's entitlement and the value of any requested increase in the Contract Price and/or the Contract Time. Claims seeking an extension in the Contract Time shall include a schedule analysis substantiating such extension. Owner may request additional documentation if in its judgment such additional documentation is reasonably necessary to determine if Contractor's Claim is justified as to entitlement and/or value.

11.5   *Owner's Decision* Owner shall issue a written decision in response to each Claim within thirty (30) days of receipt of the Claim by Owner. Such decision shall approve the Claim, deny the Claim, or approve the Claim in part and deny it in part. Prior to the expiration of such thirty (30) days, Owner may explain in writing why a decision cannot be rendered in thirty (30) days and establish a date by which such decision shall be issued.

11.6   *Notice of Intent to Litigate* If Contractor is not satisfied with Owner's decision with respect to any Claim submitted by Contractor, Contractor may, within thirty (30) days of receipt of Owner's decision, submit a notice of intent to litigate. Failure to submit such notice of intent to litigate in a timely manner shall preclude Contractor from bringing any legal action with respect to the matter covered by such Claim.

11.7   *Commencing Litigation* Contractor, following its submission of a notice of intent to litigate, may commence an action at law or in equity in the [appropriate court in the state where the project is located]. Contractor may only commence litigation if all requirements of Article 11 have been satisfied, it being the express intent of the parties that no issue arising out of or related to the performance of this Contract shall be litigated unless it has first been submitted by Contractor as a Claim and Owner has issued a written decision with respect to such Claim, provided that if Owner has not issued a written decision pursuant to Paragraph 11.5 within thirty (30) days following final completion of the Work, Contractor may then commence litigation notwithstanding that Owner has not issued a written decision.

11.8   *Alternative Dispute Resolution* Owner and Contractor may, but shall not be required to, agree to mediate and/or arbitrate any dispute prior to the filing of a legal action. Such agreement shall be in writing, signed by both parties, and shall incorporate the applicable rules of the American Arbitration Association.

## Article 12 Termination

12.1   *Termination for Cause* Owner may terminate the Contract if Contractor

a.   fails materially or persistently to perform the Work as required by the Contract Documents;

b.   refuses or otherwise fails materially or persistently to respond to instructions issued by Owner, Designer, or any other person or entity authorized by this Contract to issue instructions on Owner's behalf;

c.  refuses or is unable to commit adequate levels of labor, materials, and equipment to perform the Work within the Contract Time or is otherwise unable to complete the Work within the Contract Time for reasons for which it is responsible; or

d.  is unable to provide evidence of financial ability to complete the Work when such evidence is requested by Owner in response to a bankruptcy filing by Contractor or other evidence from which Owner may reasonably conclude that Contractor may be experiencing significant financial problems.

12.2  *Process of Termination* Owner shall, on the basis of one or more of the grounds for termination enumerated in Paragraph 12.1, send Contractor a certified letter, return receipt requested, or by any other method that documents receipt by Contractor, stating in detail each of Owner's grounds for termination of Contractor and giving Contractor seven (7) days to cure each such ground for termination. If at the end of the seven (7) days, Contractor has not cured each such ground for termination, or for those grounds which cannot be fully cured in seven (7) days, made reasonable progress in curing such grounds, Owner shall terminate Contractor by delivering in hand to Contractor at Contractor's office at the Site written notice that Owner has terminated the Contract pursuant to its letter described in this Paragraph ("Termination Notice"). Contractor shall immediately cease performing the Work, and Owner shall immediately thereafter take possession of the Site, including any of Contractor's materials and/or equipment, and shall complete the Project in whatever manner it deems most appropriate. Copies of all correspondence required by this Paragraph shall be sent to Contractor at the address specified by Paragraph 15.7 and to Contractor's surety.

12.3  *No Further Payment to Contractor* Effective as of the date of termination as established by the Termination Notice, Owner shall make no further payments to Contractor until the Work is finally complete. Upon final completion, Owner shall determine the cost to complete the Project ("Completion Cost"), which shall include the cost of completing work not performed by Contractor, the cost of completing any work started but not completed by Contractor, and the cost of correcting any defective work performed by Contractor. If the Completion Cost exceeds the balance of the Contract Price unpaid to Contractor at the time of termination, Contractor shall pay the difference to Owner. If the Completion Cost is less than the balance unpaid to Contractor at the time of termination, Owner shall pay the difference to Contractor.

12.4  *Termination for Convenience* Owner may terminate this Contract for its convenience at any time by giving Contractor at least seven (7) days written notice by certified mail, return receipt requested, or any other manner which documents receipt by Contractor. Such notice shall include instructions to Contractor concerning how to implement the termination. Upon receipt of such notice, Contractor shall faithfully perform the instructions set forth in such notice. Final payment to the Contractor following termination for convenience, which shall be requested and paid in accordance with the provisions of Article 6, shall include any reasonable additional costs solely attributable to Owner's termination for convenience but shall not include any amounts for lost revenues, lost profits, or any other consequential damages.

12.5  *Termination by Contractor* If Owner fails to make each of two consecutive payments required by Paragraph 6.3 for at least sixty (60) days, provided that Owner's failure to make such payments is not in any way due to acts or omissions of Contractor, Contractor may terminate this Contract by sending Owner a written notice by certified mail, return receipt requested, or in any other manner that documents receipt by Owner. If Owner fails to cure the nonpayment within fifteen (15) days, or to reasonably demonstrate that its failure to make payment to Contractor is based on an act or omission of Contractor, Contractor may at the end of that fifteen (15) day period cease performing work and remove its workforce and equipment, but not materials paid for by Owner, from the Site. Final payment to Contractor shall be made in accordance with Paragraph 12.4, except that the allowable additional costs shall be those solely attributable to the early termination of work by Contractor rather than termination for convenience by Owner.

## Article 13 Records

13.1 *Books and Records* Contractor shall keep for at least six (6) years after final payment, books, records, accounts, and all documents created or received by Contractor in the performance of the Work or otherwise pursuant to this Contract. Owner, with reasonable notice, shall at any time during the progress of the Work and for such six (6) years after final payment have access to and may obtain copies from Contractor of such books, records, accounts, and documents. Owner shall reimburse Contractor for the actual cost of any copies of any documents exceeding in total ten (10) pages.

## Article 14 Waiver

14.1 *Waiver* No act or failure to act by Owner shall relieve Contractor of any obligation imposed on Contractor by the Contract Documents. Any relinquishment by Owner of any right or waiver of any Contractor obligation shall be effective only if it is in writing, signed by Owner and addressed to Contractor, and shall be effective only for the specific act or omission addressed in such written waiver and not for any other similar or different act(s) or omission(s).

## Article 15 Miscellaneous Provisions

15.1 *Integration* This Contract shall represent the entire agreement between Owner and Contractor and shall supersede any prior agreements, representations, and understandings, written or oral.

15.2 *Calendar Days* References to days in this Contract shall mean calendar days unless otherwise stated. When a specified period would end on a Saturday, Sunday, or holiday, it will be deemed to end on the next business day.

15.3 *Laws of the State of* _____ The provisions of this Contract shall be interpreted according to the laws of the State of _____.

15.4 *Headings* The headings contained in this Contract are for convenience only; they are not part of the Contract and shall not define or limit the rights or responsibilities of either Owner or Contractor.

15.5 *Unenforceable Provision* If any provision of this Contract is held to be unenforceable by a court of competent jurisdiction, the remainder of the Contract shall remain in effect as if the invalid provision had not been initially included.

15.6 *Only Owner Liable* Contractor shall seek to enforce the provisions of this Contract only against Owner. No partner, officer, board member, or employee of Owner shall incur any liability to Contractor as a result of this Contract.

15.7 *Notices* Written notices required by this Contract to be sent by one party to the other shall be sent to the addresses first appearing in this Contract to the attention of the receiving party's Project Manager. Either party may change the contractual notice address by providing written notice to the other party stating the new notice address.

Owner and Contractor each execute this Contract under Seal as of the date first mentioned in the Contract to fully memorialize the agreements and mutual consideration contained in this Contract. The Contract may be executed in multiple copies, each of which shall be considered an original.

| *Owner:* | | *Contractor:* | |
|---|---|---|---|
| By: | _____ | By: | _____ |
| Title: | _____ | Title: | _____ |

# Index

acceleration 149–50
acknowledgment/agreement language 18
acts of God 150
actual costs proposal 95
actual extra costs 147
additional services: analysis 94–5; causes 88–90; consultant agreements 108; definition 92; minimization 90–4; valuation 95–7; *see also* change orders
additional services game 62
administration *see* construction contract administration; consultant agreement administration; contract administration; design agreement administration
agreement for consulting services *see* consultant agreement (sample)
agreement for design and construction administration services *see* design agreement (sample)
agreement price not to exceed a certain value 91
allegations/accusations 67–8
ambiguity 43–4
American Institute of Architects (AIA) 43
anti-waiver provisions 52
approval gap 88
approvals *see* project approvals
arbitration 10
as-built design documents 114–15
attorneys 3
authority 162

bar chart schedules 154–5
baseline schedule 132–4
baselines 85–6
benefits 24–5
books *see* documentation
budget 73–4
building information modeling (BIM) 11
business relationship 6–7, 26; contractual 139–40; pointing an unloaded gun 66; *see also* procurement relationship

cardinal change 150–1

change order analysis 168; analyze the contractual requirements 169–70; calculate compensation 173; determine entitlement 173; evaluate the contractor's factual position 171–2; negotiate appropriate values 173; prepare a formal response 174; review the contractor's change order proposal 168
change order contract provisions 142; contractor must keep working 146; damages for delay 145; executed change order covers all damages 146; notice 142–4; required documentation 144; unilateral change orders 144–5
change order cost categories 162–3; equipment costs 165; home office overhead 167; labor costs 163–4; mark-ups 165–6; material costs 164–5; subcontractor costs 165; supplemental costs 166–7
change order requirements 146–7, 149; acceleration 149–50; acts of God 150; actual costs and/or time 147; cardinal change 150–1; defective specifications 151–2; delay 152–6; denial of access; differing site conditions 156–7; direct causation 148; documentation 148–9; entitlement 147–8; extra work 157–8; improperly denied substitution 158–9; interference 158–9; performance impossible/impractical 159; problems caused by other contractors 160; wrongful termination 160
change orders 60, 62–3, 79, 141; compensation 167–8; owner defenses 161–2; *see also* claims management
changes to the program 89
claims management 60, 79, 141, 174–5, 181; *see also* change orders
clear drafting 91–3
closeout *see* contract closeout
collaboration 10–12
commissioning 180
communications *see* correspondence
compensation *see* damages
complete design services 74–5

For Product Safety Concerns and Information please contact our EU
representative GPSR@taylorandfrancis.com Taylor & Francis Verlag GmbH,
Kaufingerstraße 24, 80331 München, Germany

Printed and bound by CPI Group (UK) Ltd, Croydon, CR0 4YY

08/05/2025

01864391-0009